W9-AVB-366

The HIDDEN BOOK *in the* BIBLE

Also by Richard Elliott Friedman:

The Exile and Biblical Narrative

The Creation of Sacred Literature, Editor

The Poet and the Historian, Editor

The Future of Biblical Studies: The Hebrew Scriptures, Co-editor

Who Wrote the Bible?

The Hidden Face of God
(originally published as *The Disappearance of God*)

The HIDDEN BOOK in the BIBLE

Restored, Translated, and Introduced by

Richard Elliott Friedman

HarperSanFrancisco
A Division of HarperCollins*Publishers*

Some material in the Introduction and Appendix appeared in articles by the author in *Bible Review* and is reprinted here with permission: "Deception for Deception," *Bible Review* II:1 (1986), pp. 22–31, 68; "Late for a Very Important Date," *Bible Review* IX:6 (1993), pp. 12–16; "Scholars Face Off Over Age of Biblical Stories," *Bible Review* X:4 (1994), pp. 40–44, 54.

Designed by Interrobang Design Studio

FIRST EDITION

Library of Congress Cataloging-in-Publication Data
Friedman, Richard Elliott.
The hidden book in the Bible/restored, translated, and introduced by Richard Elliott Friedman.
Includes bibliographical references.
ISBN 0-06-063003-5 (hardcover)
ISBN 0-06-063004-3 (pbk.)
1. J document (Biblical criticism) 2. Bible. O.T. Former Prophets—Criticism, Textual. I. Title.
BS1181.4.F72 1998
222′.05209—DC21 98-22585
CIP

98 99 00 01 02 ❖ RRD(H) 10 9 8 7 6 5 4 3 2

This book is dedicated to my beloved daughter

Jesse Rebekah Friedman

who has brought me so much happiness

CONTENTS

PREFACE AND ACKNOWLEDGMENTS

I arrived at this discovery twelve years ago. Because of its potential significance, I proceeded extremely slowly and cautiously. I passed it before as many colleagues as possible to get their reactions and criticism. I first presented it as a paper before the Biblical Colloquium, the forty-year-old distinguished society of biblical scholars, at its annual meeting held at Princeton University in 1986. I then presented it to seminars at the University of Cambridge; Yale University; the University of California, Berkeley; the University of California, San Diego; the Hebrew University in Jerusalem; and the American Schools of Oriental Research at the Albright Institute in Jerusalem. The overall response was encouraging. I learned much from criticisms that were raised, and I addressed them in subsequent research, and the case became progressively clearer as a result. The enthusiasm of the response grew over the course of the presentations. I also tried out the idea at meetings of clergy and in open lectures at universities and to the public. And I used my best, time-tested means of working out a new idea: I gave a course on the material and worked through it with my students.

The twelve years that it has taken have been well worth it. I present the restored work that appears in this book with confidence that it is in fact the first great prose work of world literature. I have tried to present it in such a way that both laypersons and scholars will be able to judge the evidence for themselves and see its remarkable unity.

I am particularly grateful to my colleagues in San Diego at the University of California: David Goodblatt, Thomas Levy, William Propp, and especially our senior colleague, David Noel Freedman. It is remarkable fortune to have such a group of admirable scholars right in one's own backyard, to share and test ideas, to catch mistakes, and to encourage one another.

The editor of this book, at Harper Collins, is Mark Chimsky. The work benefited from his superb skills and understanding.

Elaine Markson, my literary agent, nurtured the work from start to finish.

And, as always, my wife, the legal anthropologist, listened patiently to things that are not in her field, and, as always, improvement resulted from her good sense.

And so I am grateful to the many readers and listeners who refined this work by their reactions, comments, and wisdom. I hope that, with their help, I have done justice to the great anonymous author who gave us this treasure.

Richard Elliott Friedman

Jerusalem

April 1998

INTRODUCTION

1. THE RECOVERY OF THE WORK

A GREAT WORK LIES EMBEDDED IN THE BIBLE, A CREATION THAT WE CAN TRACE TO A SINGLE AUTHOR. AND I BELIEVE THAT WE CAN ESTABLISH THAT IT IS OF great antiquity: it was composed nearly three thousand years ago—so it is indeed nothing less than the first work of prose. Call it the first novel if you think it is fiction, or the first history if you think it is factual. Actually, it is a merger of both. But, either way, it is the first. There is no long work of prose before this anywhere on earth, East or West, so far as I know. We know of poetry that is earlier, but this is the oldest prose literature: a long, beautiful, exciting story. And the astonishing thing is that, even though it is the earliest lengthy prose composition known to us, it is far from a rudimentary, primitive first attempt at writing. It has the kind of qualities that we find in the greatest literature the world has produced. Indeed, scholars of the Bible and of comparative literature have compared individual parts of it to Shakespeare and to Homer. Those scholars were right, but they were barely at the threshold of the full work, a composition whose unity and brilliant connections have been hidden by the editorial and canonical process that produced the Bible.

This hidden book was originally a united story, but it was cut up by the Bible's editors, and then other stories, laws, and poetry were spliced into it and around it. And so the divided pieces of this saga are now spread out through nine books of the Bible, from Genesis to the first two chapters of Kings. What I have done is to separate the original text from all of those other writings that have surrounded it. Imagine that such a discovery had been made through archaeology instead of in the less romantic setting of biblical scholarship. And imagine that you had the good fortune to be in the archaeologist's office. He goes over to the file drawer and says, "Oh, did you want to see it?" Wouldn't you want to read it immediately? That is what this book purports to be: the opportunity to read this work that no one has read for almost three millennia. If we had just discovered a work of this quality and this length archaeologically, we would be impressed and excited. But something even more impressive has happened. This masterpiece has been under our eyes, laced in the fabric of the best known book in the world, for thousands of years. And it was not just a part of the Bible. It was the heart of the Bible, the core work. The Bible's other texts were added to it and assembled around it.

Some people praise the Bible's literary artistry. Others are interested in it for history, with some believing that it is entirely historical and others weighing each of its reports critically. Still other people are most interested in the Bible for theology, trying to make sense of its conception of God—and its conception of the relationship between God and humans. The work of the first prose writer displays greatness in each of these arenas. In artistry and substance, it is among the great works of literature of all time. In terms of history, it is the first known attempt at history writing—and is impressive for such an early attempt. And, theologically, its conception of the relationship between God and humankind affected the course of virtually all subsequent ideas of God in Western religion.

Here is a story that begins "in the day that YHWH (God) made earth and skies" and ends with divine promises fulfilled and with humans taking responsibility for the land that their creator has promised them. Here is a return to the foundation, to the original conception of God and humanity—before we added layers and layers of later conceptions to it. All those layers made it a rich and complex thing. But there is something pure, spontaneous, and beautiful about the original thing. And I think that it can be enlightening and exciting for our age. For those who do not know the Bible well, this is an entry into it from its original form. For those who do know the Bible well, to read this account is to experience the feeling of being close to the Bible's heart. The past two centuries of biblical scholarship sought to uncover the process that produced the Bible. Now, after twenty years in this field, I have a new sense of where much of that scholarship was going: to bring us back to where we could *experience* that process. When those of us who know the Bible and its history well read this text, we can feel the Bible growing up around it. Critical scholarship has been used by some to chop up the Bible, to strip away its Bible-ness. But, perhaps unknown even to those who contributed to it, biblical research was going somewhere else. It was leading us back to the experience of the process of the formation of the Bible, like watching a time-lapse film of a blossoming flower.

Here is an account of the path that led me to this work:

Scholars in the last century unfolded a picture of how the first five books of the Bible came to be written. These books are known as the Torah, the Pentateuch, or the Five Books of Moses. I told the remarkable story of how scholars arrived at this understanding in an earlier book, *Who Wrote the Bible?* The method that produced this picture has come to be known as Higher Criticism, and the picture itself has come to be known as the Documentary Hypothesis. Its root idea is that the Torah was not written by one person, Moses, but was a product of several source works that were combined by editors to form the Five Books of Moses. The four largest of these source works are known classically by the symbols J, E, D, and P. The works that are known as J and E have been regarded as the earliest, written in the tenth to eighth century B.C.E.; D comes next, in the late seventh century; and P has been regarded as the latest, from the sixth or fifth century. There is still plenty of disagreement about individual elements of the model, and there are variations such as supplementary hypotheses.

I, too, have disagreed with one portion of it in particular: the idea that P is the latest work.[1] But the basic picture has become widely accepted—except in orthodox Jewish and fundamentalist Christian communities, where people believe that Moses wrote down the text. The Documentary model is taught at most universities and seminaries, and it appears in most introductions to the Bible. Priests, ministers, and rabbis are acquainted with it, and a number of books have made it known among the general public.

I think that most people who are familiar with this model would say that they like J best of the four main source works. It contains particularly human-centered versions of the stories of creation, the flood, the covenant between God and Abraham, the devastation of Sodom and Gomorrah, the marriage of Rebekah and Isaac, the struggles between the twin brothers Jacob and Esau, and more. And it contains famous stories that are uniquely J, with no parallel accounts in P, E, or D, such as the stories of Adam and Eve, Cain and Abel, the tower of Babylon, the three visitors to Abraham, the story of Dinah and Shechem, Judah and Tamar, and more. The stories in J are disproportionately among the Bible tales most enjoyed and remembered by children, and they are the most esteemed for their artistry by literary critics and Bible scholars alike. As of the time that I am writing this, there have been three books in which J has been translated into English and presented separately from the other three sources, along with literary, historical, and theological analyses.[2] The first of these books, by Peter Ellis, called the author of J "the Hebrew Homer." The last, by Harold Bloom, repeatedly compared the author to Shakespeare. I have no quarrel with any of this (except that these authors would have been more accurate if they had compared it to the works of Dostoevsky than to those of poets). I have loved these stories since I was a child; and now, as a professor of Hebrew and comparative literature, I value the J narrative among humankind's great literature. But I have found that J is not a work. It is the *begin-*

ning section of a work: a long, exquisitely connected prose composition whose artistry and power extend from its beginning in J to its finish in the story of a royal family. It is as if we had only Chapter 1 of *The Brothers Karamazov,* and we all thought it was marvelous, and then someone found the rest of the book in Dostoevsky's closet, and we suddenly realized that Chapter 1 was just the first part of a masterpiece.

The recovery of this work began when a colleague of mine at the University of California, San Diego, told me many years ago that he thought that J was composed by the same author as the section of the Bible known as the Court History of David. The Court History is the story of King David and his family. It takes up most of the book of 2 Samuel. This colleague, Jonathan Saville, was not a biblical scholar, but he was familiar with scholarship in this field and had a solid command of biblical Hebrew, and he was a sensitive and thoughtful reader. Elements of language, style, and interests in these two works seemed to him to be so close as to lead one naturally to feel them to be by the same person. The great literary critic Erich Auerbach, as well, in a classic study that several people trace as the starting-point of contemporary study of the literary artistry of the Bible, spoke of J and the Court History as coming from the same hand. Auerbach did not present an argument to prove this point; he just lightly referred to it as if it were too obvious to require demonstration.[3] I would say that Auerbach and Saville each arrived at this sense that the two works were by the same author by way of their instincts as readers, not by systematic reasoning.

Meanwhile, in the 1970s, a biblical scholar, Robert Polzin, did a study of the stages of the Hebrew language in biblical prose. Polzin showed that J and the Court History reflected the same stage of biblical Hebrew. They had to have been written in the same period.[4]

When I first heard this idea from Saville, I was not sure what to make of it. I had respect for his literary sense, and I was impressed that Auerbach had felt the same thing. I was also impressed by the growing body of work in linguistics, like that of Polzin's, that was enabling us to date biblical texts by the stage of the Hebrew language that they reflected.[5] Some observations that I had made in my own work on J further fit with this idea that there was some meaningful connection between those J stories in Genesis and the stories of David's court in the book of 2 Samuel.

In the first place there were the parallels in the stories themselves. Consider the story of Jacob in J and the story of King David in the Court History. Both show an obvious concern with the succession of sons to the place of their father. In J, four of Jacob's sons are contenders: Reuben, Simeon, Levi, and Judah. In the Court History, likewise, four of David's sons are in the running: Amnon, Absalom, Adonijah, and Solomon. And in both stories the fourth son is the victor. And the offenses that eliminate Reuben, Simeon, and Levi in J are the same offenses that eliminate Absalom from the succession in 2 Samuel: Reuben sleeps with his father Jacob's concubine; Absalom takes ten of his father David's concubines! Simeon and Levi avenge a sexual injury to their sister, Dinah, by murdering the man who did it; Absalom avenges the rape of his sister, Tamar, by murdering the man who did it. And both stories have a formerly strong, now comparatively weak father, who hears about the injury to his daughter but does not act: Jacob in J, David in the Court History.

J is disproportionately about Judah, which is King David's tribe. The covenant promise that God makes to Abraham in J is fulfilled in David in the Court History, who controls the promised territory from the River of Egypt to the Euphrates.

Both J and the Court History are about families, and the plots of both depend on chains of deceptions and recompense that run through those families. Both have wise, controlling females acting within a patriarchal structure. Both have wronged females. Both have a woman named Tamar. Both have a woman called Bathsheba (or Bath-shua).[6] Both have openly imperfect heroes. David, like Jacob, has twelve sons in Jerusalem.[7] Freudian literary analysis also reveals a range of similarities between these two groups of stories—relations of fathers and sons, sibling rivalries, powerful mothers developed outwardly as relatively minor characters—in a concentration found nowhere else in the Bible.

But these parallels, plus the linguistic evidence, just show that these two sections of the Bible were written in the same period and have some things in common. This still does not mean that both were written with the same stylus. After all, there had always been people who recognized that *something* was going on between these two texts. One writer emphasized their similarity of style and traced them to the same period.[8] One writer saw Genesis and 2 Samuel as "companion works," with some of these stories reflecting and elaborating on the stories of David.[9] One pictured an editor reworking one text to match the other.[10] Another suggested that the J author in Genesis was influenced by the life and times of King David himself, and that this was the reason for the similarities.[11] The literary critic Harold Bloom claimed that the authors of the two works were friends and rivals.[12] How does one pursue an idea like this? How does one find out whether two works that have some similarities were actually written by the same person? That was the question that I set out to answer, but, as sometimes happens in research, the path that I took to answer this question led me to something much bigger than I had set out to find.

As I made a plan to pursue these similarities between the two works, the first thing I had to do was to determine where J and the Court History each began and where each ended. In the case of J, it is easy enough to know where it begins. It begins in Gen 2:4 with an account of creation, starting with the words:

In the day that YHWH made earth and skies . . .

The question was, where does its story end? People frequently say that the story simply does not end at the death of Moses. The promise to Abraham in Genesis—that his descendants will one day be a nation that will possess the land of Canaan—does not come true until after Moses' death in Deuteronomy. This promise is fulfilled in the next book of the Bible, the book of Joshua. Moreover, the first twelve and a half chapters of Joshua have substantial similarities to J and make some specific mentions of things that happen in J, as we shall see. And so I tentatively included these parts of the book of Joshua in order to see if this would hold up.

Next came the question of where the Court History of David begins and ends. It includes nearly all of the book of 2 Samuel and appears to end with the death of David and the succession of his son, Solomon, which come in the first two chapters of 1 Kings. Determining its starting point is more difficult. Its story clearly depends upon events that occur in 1 Samuel, but how far back into 1 Samuel does it go? Scholars used to see the book of 1 Samuel as a combination of two originally separate texts, known as Samuel A and Samuel B. This classic division of the book of 1 Samuel into two large sources has been widely abandoned in recent scholarship, but I began with it as a working hypothesis. Literary factors still seemed to me to point to this division of 1 Samuel.[13] I made tentative identifications of Samuel A and B, which did not differ dramatically from past scholars' identifications.[14] In such identifications, it is manifestly the part known as Samuel B that flows into the Court History and on which the Court History depends explicitly. I shall show exactly which stories I mean in a moment.

So now I was looking at texts that were strewn from Genesis to Joshua and from 1 Samuel to Kings. The only book left between these two bodies of text was the book of Judges, and as I started the investigation the evidence showed that some of this book would figure in the puzzle as well.[15] The strange thing was that all of the evidence that connected the book of Judges to this work was in chapters 9–21. No evidence at all came from the first eight chapters. This was intriguing to me because another scholar, Baruch Halpern, had found the first eight chapters to be a separate work from the rest of the book on entirely different grounds a few years earlier.[16]

It was not my intention at the start of this research to include texts from the books of Joshua, Judges, or 1 Samuel. I simply meant this to be a study of J and the Court History of David. But, as we shall see, the evidence required the connection of these intervening texts with the others in the study. I was finding that the text that is known as J is just the beginning of a story that starts with the creation of the world in Genesis and continues all the way to the establishment of David's kingdom on the earth.

What is the right metaphor to capture this? I could compare it to the pieces of a jigsaw puzzle coming together, except that we had not even known that they *were* just pieces—or that this particular puzzle even existed. These texts, spread through nine books of the Bible, were sensible, attractive stories on their own. But when they came together, a parade of signs of their unity appeared, and a hidden book emerged. The total of all of those pieces is a work of about three thousand sentences—not long by the standard of contemporary novels but quite long by the standard of biblical sources, and extraordinarily long for the period in which it was born. For those who want to note the exact passages, here is a list of them:

Genesis 2:4b–25; **3**:1–24; **4**:1–24, 26b; **5**:28b–29; **6**:1–8; **7**:1–5, 7, 10, 12, 16b–20, 22–23; **8**:2b–3a, 6, 8–12, 13b, 20–22; **9**:18–27; **10**:8–19, 21, 23–30; **11**:1–9; **12**:1–4a, 6–20; **13**:1–5, 7–11a, 12b–18; **15**:6–12, 17–21; **16**:1–2, 4–14; **18**:1–33; **19**:1–28, 30–38; **21**:1a, 2a, 7; **22**:20–24; **24**:1–67; **25**:8a, 11b, 21–34; **26**:1–33; **27**:1–45; **28**:10, 11a, 13–16, 19; **29**:1–35; **30**:24b–43; **31**:3, 17; **32**:4–14a; **33**:1a, 3b, 4, 16; **34**:1–31; **35**:21–22; **36**:31–39; **37**:2b, 3b, 5–11, 19–20, 23, 25b–27, 28b, 31–35; **38**:1–30; **39**:1–23; **42**:1–4, 6, 8–20, 26–34, 38; **43**:1–13, 15–23a, 24–34; **44**:1–34; **45**:1–2, 4–28; **46**:5b, 28–34; **47**:1–6, 27a, 29–31; (**49**:1–27); **50**:1–11, 14, 22b;

Exodus 1:6, 22; **2**:1–23a; **3**:2–4a, 5, 7–8; **4**:19–20a, 24–26; **5**:1–2; **13**:21–22; **14**:5a, 6, 9a, 10bα, 13–14, 19b, 20b, 21b, 24, 25b, 27b, 30–31; **16**:4–5, 35b; **19**:10–16a, 18, 20–25; **34**:1a, 2–28;

Numbers 10:29–36; **13**:17–20, 22–24, 27–31, 33; **14**:1b, 4, 11–25, 39–45; **16**:1b, 2a, 12–14, 25, 27b–32a, 33–34; **20**:14–21; **21**:1–3, 21–35; **25**:1–5;

Deuteronomy 34:5–7;

Joshua 1:1, 6, 9b; **2**; **3**; **4**:14; **5**:1–9, 13–15; **6**; **7**; **8**:1–29; **9**:1–15a, 16, 22–26a; **10**; **11**:1–23; **13**:1–13;

Judges 8:30–32; **9**:1–16, 18–57; **10**:8–9, 17–18; **11**; **13**:2–25; **14**; **15**:1–19; **16**:1–31a; **17**; **18**; **19**; **20**; **21**;

1 Samuel 1; **2**; **3**; **4**:1–18a, 19–22; **5**; **6**; **7**:1–2, 5–17; **8**:1–3a, 4–7, 9b–22a; **10**:17–27; **11**:1–15; **12**:1–6, 16–20a, 22–23; **15**; **16**; **17**:1–11, 50; **18**:6–13, 16, 20–21a, 22–29; **19**:1–24; **25**; **26**; **27**; **28**; **29**; **30**:1–6, 8–31; **31**:1, 8–13;

2 Samuel 1; **2**:1–9, 10b, 12–32; **3**; **4**; **5**; **6**; **7**:1a, 2–11a, 11c–12, 18–21, 25–29; **8**; **9**; **10**; **11**; **12**; **13**; **14**; **15**; **16**; **17**; **18**; **19**; **20**:1–23; **21**;

1 Kings 1:1–53; **2**:1–2, 5–46

When we separate these texts from the rest of the Bible and read them in order, we find things that appear regularly in these texts but *nowhere else* in the Bible's narrative. We find words and phrases that occur all through this group of texts but not elsewhere. We find that they tell a flowing, continuous story. We find exquisite connections showing that they originally went together: quotations, allusions, puns, and a variety of literary devices that fine writers of every period have used to bind works together. And I found that all these connections were not the work of editors who inserted them in order to make various short texts fit together better. On the contrary, it was precisely the Bible's editors who broke the unity and continuity of this work by mixing other works into it. If you ask, "Why have we never seen this before?" I answer that it is because the editors of the Bible were brilliant. At least three editors added new works to this one in a process so sophisticated and complicated that it has taken us centuries to see it and unravel it. We came to see it in stages, as many scholars saw parts but not all of it. Auerbach saw the connection between J and the Court History. The German scholar Budde probably came the closest over a hundred years ago, seeing J as extending from Genesis all the way to 1 Kings 2, but he did not attribute it to a single author and his work has rarely been followed.

Particular words and phrases occur commonly in this group of texts but nowhere else in biblical prose. These expressions cross many lines of genre, theme, and subject matter, so we cannot explain their recurrence as simply reflecting some common themes in these texts. In a story from this group, Jacob says to his father-in-law, Laban, "Why did you deceive me?" In another story, Joshua says the same words to the people of Gibeon. In another story, Saul says it to his daughter, Michal. In another, the mysterious woman of En-dor says it to King Saul. Of five occurrences of this expression for deception and two occurrences of the related Hebrew phrase "with deception" in biblical prose, all seven are in this group of texts. Both cases of a "coat of many colors" are in this group and nowhere else in biblical prose. The expression "flesh and bone" occurs seven times in biblical prose, and six are in this group. The expression "kindness and faithfulness" occurs seven times in biblical prose, and all are in this group.[17] Many people are familiar with a common image in the Bible in which a host washes his guests' feet. There are seven occurrences of the expression "to wash the feet" in biblical prose. All are in this group. Many are familiar with the biblical term "Sheol," referring to some not-yet-understood place of the dead. *All* references to Sheol—occurring nine times in biblical prose—are in this group of texts.

All nine references to shearing sheep are in this group.

The term for a foolish person or thing *(n^e^bālâ)* occurs ten times in all of biblical prose. All ten are in this group of texts.

The term "to lie with" with sexual connotation occurs thirty-two times in biblical prose. Thirty of them are in this group.[18]

Now, I recognize that in a group of texts this large we should expect a certain number of cases of overlap like these. But the number we are observing here is substantial, and this is only a sampling of the cases. This list goes on and on. A chart of the evidence appears in the Appendix, pp. 379–387. We would expect that more of these terms would turn up in all of the rest of biblical prose: in E, in P, in the second half of Joshua, the first eight chapters of Judges, Samuel A, 1 Kings, 2 Kings, Ruth, Esther, 1 Chronicles, 2 Chronicles, Ezra, Nehemiah, and Daniel. But they do not. These words and expressions are distributed through J but not through E, P, or D; through Samuel B but not Samuel A; through Judges 9–21 but not Judges 1–8. They form an interlocking pattern of terminology through this particular group of texts. In statistical terms, this group is about 25 percent of the Bible's prose, but, as even a first glance at the chart will tell you, the number of cases of 90 or 100 percent of the occurrences of a term happening in this group is considerable.

If there had been nothing but this evidence of distribution of words, it would already have been interesting and suggestive, but there was more than this. Once I saw these patterns of language that fell only in certain texts, I found that there were other clues that linked the very texts in which this distribution of words and phrases occurred to each other. These clues showed that the appearances of these characteristic words was not a matter of chance but, rather, reflected a literary relationship.

First, I found that the very texts in which these chains of common wording occur connect to each other in order. When we separate this group of texts and read it through, we discover that it is more than a collection of stories with a great amount of common language. It is a continuous account. Where one text leaves off, the next text that has some of the common wording picks up the story. So, for example, in the last J passages in the Pentateuch, the people of Israel are located at a place called Shittim (Num 25:1); and Shittim is where they are located at the beginning of the book of Joshua at the point where the common language begins (Josh 2:1). Likewise, the conclusion of Judges connects to the beginning of Samuel B, the next text that has the familiar wordings. And Samuel B then flows integrally into the Court History, which depends on it for the introduction into the narrative of a number of the central persons in the story, including King David's wives, his generals, and his priests. The end of each section thus connects to the beginning of the next (with one exception, which I'll discuss later).

Second, sometimes the density of these overlapping terms in specific stories is far too great for explanation by simple chance or expected patterns of distribution. Compare, for example, the two stories I mentioned earlier of brothers avenging a sexual violation of their sister: the J story of Dinah and Shechem (found in Genesis 34) and the Court History story of Amnon and Tamar (found in 2 Samuel 13). Look at the clusters of language in the two stories:

In 2 Samuel, Amnon takes Tamar, "and he degraded her and he lay with her."

(2 Sam 13:14)

In Genesis, Shechem takes Dinah, "and he lay with her and he degraded her."

(Gen 34:2)

—~—

In 2 Samuel, Tamar tells Amnon, "Such a thing is not done in Israel" and "Don't do this foolhardy thing."

(2 Sam 13:12)

In Genesis, Dinah's brothers are upset because Shechem "had done a foolhardy thing in Israel . . . and such a thing is not done."

(Gen 34:7)

—~—

Tamar says that it would be a "disgrace" for her.

(2 Sam 13:13)

Dinah's brothers say that mixing with the uncircumcised men of Shechem's city would be a "disgrace."

(Gen 34:14)

—~—

Absalom tells Tamar, "Keep quiet."

(2 Sam 13:20)

Dinah's father, Jacob, "kept quiet."

(Gen 34:5)

—~—

The man who degrades Tamar dies violently at the hands of her brother.

The man who degrades Dinah dies violently at the hands of her brothers.

Tamar's father, David, knows but is passive, his son takes vengeance, and he is angry afterward.

(2 Sam 13:21)

Dinah's father, Jacob, knows but is passive, his sons take vengeance, and he is angry afterward.

(Gen 34:30)

The parallels to the Court History's Amnon-and-Tamar episode come elsewhere in J as well, not just in the Dinah-and-Shechem story. As I have noted, J has a Tamar also, the ancestor of the latter Tamar. Both are stories about sexual relations within a family. Revenge is taken for the Court History's Tamar when they are shearing (2 Sam 13:24); revenge is taken for J's Tamar when they are shearing (Gen 38:12f.). In the Court History, Tamar, the innocent victim of violence by her brother, wears a "coat of many colors" which is torn (2 Sam 13:18f.); in J, Joseph, the innocent victim of violence by his brothers, wears a "coat of many colors" which is torn (Gen 37:3, 23, 32). (And recall that these are the only occurrences of the coat of many colors in the Hebrew Bible.) In the Court History, David "mourned over his son all the days" (2 Sam 13:37). In J, Jacob "mourned over his son many days" (Gen 37:34).

Here is an even more striking case: In the famous J story of Sodom and Gomorrah, two travelers (who happen to be angels) arrive in Sodom. Lot, who is Abraham's nephew, shows them hospitality, but the people of Sodom surround the house and demand that he send the guests out to the crowd. In a story that is found in the text that I identified in the book of Judges, some travelers (a man and his concubine) arrive in a city in Benjamin. One man shows them hospitality, but the people of the city surround the house and demand that he send the guest out to the crowd.

In Genesis Lot says to the angels, "Turn . . . and spend the night" (Gen 19:2). In Judges the travelers "turned to spend the night" (Judg 19:15).

In Genesis the angels answer, "We'll spend the night in the square" (v. 2). In Judges the old man says, "Don't spend the night in the square" (v. 20).

In Genesis Lot "pressed" (Hebrew root *psr*) the men to spend the night (v. 3). In Judges the concubine's father "pressed" *(psr)* his son-in-law to spend the night (v. 7) .

Genesis says, "and they came to his house" (v. 3). Judges says, "and he had him come to his house" (v. 21).

In Genesis Lot offers the visitors the washing of feet (v. 2). In Judges, too, "they washed their feet" (v. 21).

In Genesis, "the people of the city surrounded the house" (v. 4). In Judges, "the people of the city surrounded the house" (v. 22).

The people of Sodom tell Lot, "Bring them out to us, and let's know them!" (v. 5). In Judges, the people say, "Bring out the man . . . and let's know him!" (v. 22).

In Genesis Lot goes out to talk to the crowd: "And Lot went out to them" (v. 6). In Judges the old man goes out to the crowd: "And the man went out to them" (v. 23).

Lot pleads with the crowd: "Don't do bad, my brothers" (v. 7). The old man pleads: "Don't, my brothers. Don't do bad" (v. 23).

Lot offers his virgin daughters to the crowd (v. 8). The old man offers his virgin daughter to the crowd (v. 24).

Lot "delays" (v. 16).[19] The man and the concubine in Judges "delay" (v. 8).

It should be getting obvious that there is something going on here that is more than an editor with an eraser and a pencil. And there is further evidence that it is something *pervasive* that is happening. There are at least five sets of these parallel stories that have dense clusters of common terminology spread through this connected group of texts. They occur all across the group: in J, Joshua, Judges, 1 Samuel, and 2 Samuel. In the past, we have explained such parallels by saying that one author imitated another, or that one author was influenced by the actual history reflected in the other story, or that both authors used common formulas from old oral traditions, or that an editor reconciled the two stories. But none of these solutions will work when we take all of the other evidence into account as well. All five of these sister stories were part of a continuous, connected history. And that history repeatedly used words and phrases that occurred nowhere else.

There were still more clues that all of these texts had always belonged together. The clues were the kinds of things we find in any coherent work of literature. There are connecting elements of the story running through this collection. For example, the ark containing the tablets of the Ten Commandments is a continuing narrative concern in this collection. The ark is referred to in J (Num 10:33, 35; 14:44). Its importance continues to be developed in the Joshua passages (chaps. 3; 4; 6; 7; 8). And it is mentioned in a passage in one of the identified chapters of Judges (20:27), in many places in Samuel B (in 1 Samuel 3; 4; 5; 6; 7), and several times in the Court History (2 Sam 6; [7:2]; 11:11; 15:24, 25, 29; 1 Kgs 2:26). The connection is italicized by a shared prose image: when the ark is carried around Jericho in the book of Joshua, the people "shouted a big shout" (Josh 6:5, 20); and when the ark is carried into an ill-fated battle against the Philistines in Samuel B, the people "shouted a big shout" (1 Sam 4:5).

One might think that something as important as the ark is simply a common element of biblical stories, but this is not in fact the case. The ark appears in J but *never* in E. It is mentioned many times in

Samuel B, but only once in Samuel A. It occurs at the climax of the story in 1 Kings 2, but never in the books of 1 Kings or 2 Kings following the report of its being placed in the Temple (1 Kings 8).[20]

For another example of such narrative continuity there is the matter of giants. There are huge, frightening persons in the Bible. Their origin is explained in a story near the beginning of the collection: "sons of God" (whatever that means) have relations with human women, and they give birth to giants (Gen 6:1–4). Later, when Moses sends spies to scout the land of Canaan, the Israelite spies see the giants (Numbers 13). Then, in the book of Joshua, Joshua eliminates the giants from all of the land except from the city of *Gath* and two other Philistine cities (Josh 11:21–22). And then, in the Samuel B part of the collection (1 Sam 17:4), the most famous Philistine giant, Goliath, comes from that city: *Gath!*

There are many other elements of the story that thus continue through this group of texts *but not through the rest of biblical narrative.* I think that you will observe many of them as you read the work. I shall refer to more of them in my literary treatment of the work below. I believe it will become clear that this is a connected group of texts, telling a continuous, coherent story.

As one reads it, one comes across *allusions* to details that came earlier in the story. For example, in the J story of the spies whom Moses sends to scout the promised land, the people are frightened by the spies' report of the giants and the fortified Canaanite cities. They say, "Let's go back to Egypt!" (Num 14:4). In the face of their rebellion and lack of trust, YHWH (which is the way that the name of God is written in this work) swears that they will not live to arrive in the land:

> "I swear that all of these people, who . . . haven't listened to my voice, won't see the land that I swore to their fathers"
>
> (Num 14:22f.)[21]

Two books later, one of the passages in Joshua alludes to that episode of the spies in Numbers, and it refers explicitly to the words of the J version of that story. It refers back to the nation

> "who had not listened to YHWH's voice, to whom YHWH had sworn that He would not have them see the land that YHWH had sworn to their fathers"
>
> (Josh 5:6)

Indeed, there are whole chains of such allusions binding this work together. I referred earlier to the episode in which Simeon and Levi kill Shechem and his father, Hamor, over the sexual abuse of their sister, Dinah. Shechem is the prince of a city that is also called Shechem. This takes place in a J story in Genesis. There is an explicit reference back to this in the middle of a story that takes place six books later, in a passage in Judges that is part of this work. The Judges text is about a man named Abimelek, the first Israelite to attempt to be king. He reigns only in the city of Shechem, and the text refers to the people of that city as "the people of Hamor, father of Shechem" (Judg 9:28). Abimelek dies a wretched death when a woman heaves a stone on him from a city wall. And then, two books later, there is, in turn, an allusion to Abimelek's ignominious demise in the middle of the story of David and Bathsheba in the Court History (2 Sam 11:21). These allusions are widely separated in the present arrangement of the Bible, but they become a much tighter linkage when we separate this continuous strand of narrative from the mass of surrounding works. The more one reads the narrative, the more one spots these connections, and the more one senses the unity of the work.

Another body of evidence is the presence of numerous cases of repeated prose images. A scene that is pictured in one story recurs in another part of the collection, along with some recurring word or phrase to confirm that the parallel pictures are related. The J account of Moses at the burning bush includes the famous instruction, "Take off your shoes from your feet, because the place on which you're standing: it's holy ground" (Exod 3:5). And in the account of Joshua at Jericho that comes in these texts, a divine creature says the exact same words to Joshua (Josh 5:15). Both Samson (in Judges) and Samuel (in Samuel B) receive the instruction not to cut their hair; and the wording in both is "and a razor shall not go up on his head" (Judg 13:5; 16:17; 1 Sam 1:11). Both the deathbed scene of Jacob in Genesis and the deathbed scene of David in the Court History picture the father bowing on his bed to his own son (Gen 47:31; 1 Kings 1:47). The J story of Shechem (Gen 34:24f.) and the Joshua story of the Israelites' arrival in the land (Josh 5:8) both depict cases of men sitting and healing after circumcision; and both refer to the absence of circumcision as "a disgrace" (Hebrew *ḥerpâ,* Gen 34:14; Josh 5:9). There are two scenes in which women enable men to escape when they are being sought by a king's guards—a prostitute, Rahab, thus saves two spies whom Joshua sends into Jericho; and David's wife Michal thus saves him from King Saul in the Court History—and both women are pictured as saving the men in the same manner and in the same wording: "And she let them down by a rope through the window" (Josh 2:15; 1 Sam 19:12).

The issue here is more than just some similarity of wording. It is the redeployment of prose images. The parallels in the wording serve to indicate that these similar images were consciously selected. They are not just cases of an old image known from oral narrative being used in two different places. They are part of a designed construction. When one encounters them in the text, they do not come across as senseless repetitions; rather, they call to mind (sometimes consciously in the reader, sometimes not) the earlier episode and cry out for us to make the association. So when we read that a divine being tells

Joshua, "Take off your shoes from your feet, because the place on which you're standing: it's holy ground," we need not think, "That's not very original. That's the same thing that God said to Moses at the burning bush." And we cannot think that this is two authors who are just both using some famous old line, because that does not account for all the other evidence connecting the texts in which these two lines occur. We rather understand that Joshua's role as Moses' successor is being developed.

Another line of evidence further bolstered my perception that there was a conscious process in the relationship of these texts and not just recurring language and images from oral storytelling. It was the evidence of technique. First, the art of punning—paronomasia—is employed in the fabric of most of the texts in this collection. J is rich in puns of various sorts, and the same sorts of paronomasia occur in other texts from this collection as well. Now, I realize that puns occur in many prose and poetic works in the Bible, but we have to take account of them in the evidence for two reasons. First, as a general observation, the presence of puns of various sorts in so many of the texts of this collection is one more element of commonality among them. Second, and more specifically, the puns play on some of the very words and phrases that keep recurring in this collection but nowhere else. That is important because it is further evidence that the author *consciously* chose and cultivated these terms in these works. People sometimes say that puns are the lowest form of humor, but that is only when somebody else thinks of the pun first. Puns are not just cute and not even just artful. The punning on the very terms that recur in this collection reveals an author who was thinking about these words, deploying them with literary purpose. Their disproportionate occurrence in this collection is not just a matter of random distribution.

Besides punning, there were other matters of technique in this collection as well. A second one was parable. There are three uses of parables in the Hebrew Bible, and all three turned up in this collection (Judg 9:8–15; 2 Sam 12:1–4; 14:5–7). Third, people have often noted that J is rich in the use of irony, but shortly I shall show the remarkable way in which J and the Court History both use the same forms of irony to make the same point in their respective stories. Fourth, there is an uncommon degree of character development in this collection of texts. Jacob and Joseph in J are among the few figures in all biblical narrative to go through change in the course of the accounts about them. Neither of them is the same man at the end of the story that he was at the beginning. There is no such development of them in E or P. David, most patently of all, grows up and changes before our eyes in the Court History. We might possibly say this of Moses in J, and of Samson in Judges, and of Saul in Samuel B as well. There are few comparable examples of character development in all the rest of biblical narrative, where individuals tend to be essentially the same personality from beginning to end. The closest cases of such development are Moses in E and Saul in Samuel A. Not Ruth nor Esther, Elijah, Elisha, Ezra, Nehemiah, Daniel, anyone else in E, nor any king in the books of Kings or Chronicles nearly matches this degree of development.

Thematic evidence converged with all of this as well. Certain themes continue to surface—and not in a haphazard way. For example, the collection develops the matter of drunkenness five times: Noah makes wine and gets drunk after the flood (Gen 9:21); Joseph's brothers get drunk when they mistakenly think that they are out of trouble in Egypt (Gen 43:34); the priest Eli sees Samuel's mother, Hannah, praying silently and mistakenly thinks that she is drunk (1 Sam 1:13f.); Abigail's husband, Nabal, gets drunk while his wife is secretly meeting with David (1 Sam 25:36); and David later gets Bathsheba's husband, Uriah, drunk as part of David's plan to cover up his affair with Bathsheba (2 Sam 11:13). The indicator that these are not merely five unrelated tales that involve a little alcohol is the fact that in every one of these cases someone is deceived.

Similarly, the element of espionage comes up in nine different episodes spread through six books. The collection repeatedly weaves in stories of spies: Moses sends spies to scout the promised land (Numbers 14); he later sends a second set of spies (Num 21:32); Joshua sends spies to Jericho (Joshua 2); he later sends a second set of spies (Josh 7:2); spies are sent to find a territory for the tribe of Dan (Judg 18:2, 14); David sends spies to report on King Saul's movements (1 Sam 26:4); and David's son Absalom sends spies to prepare his rebellion against his father (2 Sam 15:10). In another story, the King of Ammon suspects some ambassadors whom David has sent of being spies, which sets a war into motion between Israel and Ammon (2 Sam 10:3). And Joseph repeatedly accuses his brothers of being spies in Egypt (Gen 42:9–16).

Most remarkable is the development of sexual matters as a theme in these stories. Simply put, nearly all of the sex in the Bible's prose appears in this collection. Here you will find Adam and Eve, the "sons of God" having relations with human women, Ham exposing his father Noah's nudity, two stories in which a patriarch says his wife is his sister and a king then takes his wife,[22] the angels at Sodom whom the people want to "know," Lot having relations with both of his daughters, and Jacob's wedding to Leah and then Rachel. You will read the story of Shechem's degrading of Dinah and its horrible aftermath, of Judah's unwitting sex with his daughter-in-law Tamar, of Potiphar's wife's attempted seduction of Joseph and then her claim that he tried to rape her, the Israelite men and Moabite women at Baal Peor, Rahab the prostitute, Jephthah's being identified as the son of a prostitute, Samson and a prostitute at Gaza, the raped concubine in Judges, and the priest

Eli's sons who seduce women at the Tabernacle. And you will read the accounts of David bringing a dowry of two hundred Philistine foreskins for his marriage to the princess Michal, Abner sleeping with King Ishbaal's concubine, David's self-exposing dance and his wife Michal's reaction and subsequent childlessness, David's adultery with Bathsheba, and Amnon's rape of his half sister, Tamar.

Three times this collection develops the matter of an aged man and sexual potency: in one case it is Abraham and Sarah, who doubt their ability to have a child even after hearing from God that they will have a son (Gen 18:9–14); and in the second case it is the story of David, whose rule as king ends when he is unable to have intercourse with the beautiful Abishag (1 Kgs 1:1–4). And there is the impressive description of Moses, relating that, even when he was a hundred twenty years old "his vigor had not fled" (Deut 34:7).

It twice develops the matter of a father giving his daughter, who is the hero's wife, to someone else: in the case of Samson (Judg 14:20) and the case of David and Michal (1 Sam 25:44; 2 Sam 3:13–14).

Three times it develops the matter of sons taking their father's consorts: Reuben takes Jacob's concubine Bilhah, Absalom takes ten of David's concubines, and Adonijah asks for Abishag.

And the exception that proves the rule: the people of Israel are forbidden sex for three days prior to the great revelation at Mount Sinai (Exod 19:15).

Another recurring theme: This collection repeatedly depicts the dependent position of women functioning in patriarchal society, starting with God's declaration in Eden to the first woman: "your desire will be for your man, and he'll dominate you," and proceeding to the cases of Sarah and Hagar, Rebekah, Jacob's wives and concubines, the concubine in Judges, Dinah, Tamar, Jephthah's daughter, Hannah, Samson's mother, Abigail, Michal, Bathsheba, Tamar, and Abishag.

On all of these counts—the language, the continuity, the connections, allusions, and quotations, the similarity of entire sections, the repeated prose images, the consistency of technique, and the recurring themes—the evidence indicates that these are a related body of texts.

What are we to make of all of this?

Is it chance? Given the quantity of cases, that is beyond belief.

Can the recurring images be the result of common cultural or oral background? Hardly. The concentration is too deep. The images line up in this corpus—whereas there are only a few of these with parallels in E or P (or other narrative). And there are the elements of continuity and allusion that unite this body of text in a way that common oral roots do not explain.

Is it a "school," drawing on a group of favorite images and terms? That is not likely. Usually when we speak of schools in ancient Israel we mean groups of priests or scribes who represent particular religious and social ideologies. This is what we mean when we speak of a "Priestly school" or a "Deuteronomistic school." Most of the common language that pervades the works that scholars identify as "Priestly" or "Deuteronomistic" develops the religious concerns of these works. But by far the majority of the terms and images that we have observed here do not serve such an overarching ideological purpose. They are more about sex and spies and siblings, and less about altars and law. They involve puns, irony, and allusion, but no law except the Ten Commandments. They are the images of a literary artist, not of a movement.

Is it, then, an artistic school, with the literary master being emulated by a group of disciples? That is possible, but there is a degree of continuity, coherence, direct dependence, and allusion through this procession of texts that indicates more of an overall design.

Is it, then, an artistic school work, designed by a master and filled out by the disciples—in the way that Dumas outlined his novels and then had his students write the chapters? That, too, is possible, but what evidence is there to suggest the presence of such a process here or the existence of such a school in ancient Israel? Nothing in the evidence directs us to so rare, idiosyncratic, and, at this stage of literature, unlikely a scenario.

The simple, most likely explanation, it seems to me, is that it is all by the same author. If it is not one author's work, then it is by a mysterious group of writers who all chose to use the same terms, the same idioms, the same images, and the same literary techniques and to develop the same themes—while all other biblical prose writers minimalized or rejected these terms, images, and so on. And this mysterious group of writers' individual works coincidentally connect into a continuous story with virtually no breaks except where they have been visibly combined with other works by later editors. And these writers frequently allude back and predict forward into each other's works. And they depend on each other's works for the introduction of necessary characters and events.

As radical as it may seem in the context of critical biblical scholarship—which has been *dividing* texts into ever smaller units for the last two centuries—it seems to me that the idea of one author of this work is the most natural, parsimonious, likely explanation of the evidence.

Add to this two more general considerations: First, we would be hard-pressed to find any fundamental contradictions within this work. I mean the sort of problems in the text that led scholars to divide the biblical works into sources in the first place, such as contradictions of fact, doublets of stories, conflicts of ideology or theology. The blatant, mutually supporting textual difficulties that drew scholars to think that there were a J and an E and a P and a D in the Five Books of Moses do not occur in this work.

And, second, there is a limit to just how many literary geniuses we can ascribe to ancient Israel. In biblical scholarship we have been asserting a scenario in which a very small country produced an unparalleled number of exceptional authors—and each one of them composed only one short surviving work. Again, it seems to me that one-author/one-work is still the more likely explanation.

It may have been a continuous work, now preserved almost entirely except for what the editors of the Torah cut from J. Or there may be a more complex history of composition, albeit by a single author. One possibility, for example, is that the author originally composed two works:

1. A work that contained all of the text that is now embedded in the books of Genesis through Joshua 13. This would have been a sequence from the creation and the patriarchal promises to the secure establishment of the land, perfectly understandably mentioning last the establishment of Hebron as the capital of Judah and possibly ending with the words: "And the land had respite from war."
2. A work that contained all of the text that is now embedded in the books of Judges through 1 Kings 2. This would have been a perfectly understandable sequence as a history of the establishment of the monarchy in Israel, beginning with Abimelek's failed attempt to be the first king in Israel and ending with the words: "And the kingdom was secure in Solomon's hand."

Two points of evidence support this picture. The first is the key matter of the name of the deity. The deity is always identified by name, YHWH, in J's narration, never as God (Hebrew *'ĕlōhîm*). (Individual persons in this story use the word "God" in quotation, but the narrator never uses the word.) This is true for all of the books of the Torah, from Genesis to Deuteronomy. If we look at Joshua we find that this holds true for that book as well: the word God *('ĕlōhîm)* never occurs in narration in all of the texts that fall into this work, though it does occur elsewhere in Joshua. But when we look at Judges and Samuel we find that the divine name distinction is not maintained. Now, we must be cautious on this matter, because it appears that scribes were more careful about copying the divine name in the Torah than in the other biblical books. Nonetheless, the fact that this phenomenon happens so consistently in Genesis through Joshua but not in Judges through 1 Kings 2 suggests that these may have been two separate works and that the author chose to develop the matter of the divine name in one work and not the other. Why? One can think of many possibilities. The simplest may be that the author composed the Judges–1 Kings 2 work first and only thought of making a point of the divine name matter when he or she later began the Genesis–Joshua work.

The second point of evidence is that the weakest link in the continuity of this collection comes at the same point that this divine name distinction ceases, namely, between Joshua and Judges. As I said earlier, the story in J continues logically and geographically into the beginning of Joshua, where the people of Israel are located at Shittim and are about to enter the promised land. And, likewise, the story in Judges continues logically and geographically into Samuel B, which in turn continues into the Court History. But the one juncture at which there is no such clear transition is between the end of these texts in Joshua and the point where they begin in Judges. It is therefore possible that these were originally two discrete but related works. It is striking that both works include an era of wars but end in an age of peace.

Whether this was originally one work, or two that very early were combined to make one, the fact is that either way we have before us an author's product, a lengthy work of prose, coherent and united.

One may object that the similarity of the repeated prose images, noted above, in fact argues against the idea that this was all by one author. Why, after all, would an author who was able to fashion such fine stories, with such control of technique, still be so seemingly unimaginative in using the same images repeatedly? The objection does not apply here, however, because we already knew that this author does this sort of thing, for this author does it within the J text itself. For example, the story of Jacob's meeting his wife Rachel at a well bears obvious similarity to the story of Moses' meeting his wife, Zipporah, at a well—and both of these stories are in J. Likewise the elements of the story of Abraham's telling Pharaoh that Sarah is Abraham's sister recur later in J in the story of Isaac's telling Abimelek that Rebekah is Isaac's sister. These cases are not "doublets"—i.e., two versions of the same story, involving the same persons, which suggest by their duplications of facts that they are by separate authors. These, rather, are redeployments of narrative elements—using common points in the contexts of two different stories, with each story involving different persons. And thus the full collection of texts here is no more or less likely to be by one author than is J itself. Authors are known to do this very thing, and it most certainly is not unimaginative. Milan Kundera says that after writing *The Unbearable Lightness of Being* he observed numerous occurrences of "lightness" in his earlier works. He concluded, "Only when I reread my books in translation did I see, with consternation, all those recurrences! Then I consoled myself; perhaps all novelists ever do is write a kind of theme (the first novel) and variations." Similarly I recall from my student days that after reading five of Hermann Hesse's novels *(Damian, Siddhartha, Steppenwolf, Narcissus and Goldmund,* and *Magister Ludi)* I realized that all of them were about two men, friends, whose lives part at one point, one taking a worldly path and one taking a spiritual path, and who are reunited later and share some manner of explicit or symbolic homosexual love, and then one dies. In Kundera's terms, Hesse, consciously or not, was writing variations on a theme—which need not be less valuable in literature than they are in music. Whatever our artistic judgment of such literary variations, the fact remains that authors do this, and the presence of such variations is no objection to the idea that the corpus in question is by one author.

There are many implications for biblical scholarship if this analysis is correct. It goes against the bias that biblical authors were not able to write anything long. It goes against a current view that nearly all of the Hebrew Bible's narrative was composed very late—some five hundred years or more after the latest events mentioned in this work. That view has been gaining support in recent years, asserting that the history of Israel was written by latecomers who had few sources and no idea of what had happened. The identification now of this lengthy, early work is hard evidence that that view is wrong. Also, by merging the kind of evidence that we have been using in biblical scholarship of the last two hundred years with the new literary study of the Bible of recent years, this analysis reveals how much we all gain from the new attention to the Bible's artistry. I shall take up some of these and other implications in an Appendix at the end of this book that presents the evidence for identifying this work in greater detail. To me, though, beyond any implications of this analysis is the fact that we have the work itself, a major, lengthy work by a great artist, to enjoy, to appreciate, and to study literarily, historically, linguistically, and theologically. We can look for schools, circles, movements, or chains of imitation. In fact, we have the obligation to test these possibilities. But I predict that they will not prove fruitful. The literary sensitivities of Auerbach and Saville reflected real patterns and bonds in these texts, bonds that point from several angles to the artistry of a single author—who thus would be no less than the first great writer of prose in history.

2. READING THE WORK

THOUGH I BELIEVE THAT THE MARSHALING OF EVIDENCE THAT COMES IN THE APPENDIX IS NECESSARY TO MAKE THE CASE PROPERLY FOR THE UNITY of the work, at the same time I think that it is the reading of the text that will persuade most readers of its consistency and original coherence. Indeed there is another factor that has hidden that coherence that I have not mentioned until now: translation. It is not just that editors inserted other biblical texts in between the parts of this work. There is also the fact that most translations of the Bible have been done by committees rather than individuals. Often, the Court History in the book of 2 Samuel is translated by different persons from those who translate Genesis or Judges. The result is that they choose different ways of translating a single Hebrew word. The translation of the work that appears in this book was done entirely by me, and I deliberately was conservative in my word choices, trying as hard as possible to translate consistently throughout the text, thus conveying rather than hiding or even ruining the consistent feeling of the original. (A description of the governing principles of the translation appears on pp. 57–65.) I hope that in this translation you will get a clear feeling that you are reading a construction of stories that belong together.

The thing to observe is that, when one starts to read, it seems like a sequence of loosely connected—and at times unconnected—stories. But the more one reads, and the more carefully one reads, the more one sees the network of connections through the whole text. What happens in one story has implications for things that happen in later ones. Later stories contain echoes of early stories. Later stories *allude* to early stories. Later stories *depend* on early stories. It is not the kind of similarity that stories have as a result of common roots or as a result of one account imitating another. It is, rather, the kind of structure that is conceived and executed by design. It is an elaborate embroidery of connections that pervade the entire work, the coherent production of a single mind. We can reveal this embroidery by starting with readily visible links between passages that are close together and then moving to more pervasive connections, thus:

At the first level, we see this author's techniques of connection that are obvious in passages that are nearby one another. For example, when Jacob deceives his father in order to get the blessing that was intended for his brother, Esau, Jacob puts skins "on the smooth part of his *neck*" (Gen 27:16). Later in the chapter, Esau's compensation for what Jacob has taken from him is that one day "you'll break his yoke from your *neck*" (27:40). And the story's denouement, when Esau seemingly forgives Jacob, conveys this by reporting that Esau runs and embraces Jacob and "fell on his *neck*" (33:4). The repetition of the word for "neck" forms an ironic link running through the story.

We can then move to observe such connections in passages that are more widely separated but still within Genesis. For example, when the first male is drawn to the first female, we are informed that "on account of this a man . . . *clings* to his woman" (Gen 2:24). Later comes the story in which Shechem is intimate with Dinah. Whether it is a rape, a seduction, or some other inappropriate sexual conduct, Shechem then finds afterward that "his soul *clung* to Dinah" (34:3). The phenomenon that is formulated at Eden becomes the necessary motivation of the action and outcome at Shechem.

Next we can find such connections surfacing in more distant passages. Thus, in the very same passage that tells us that "on account of this a man . . . *clings* to his woman," we also find the first man referring to the woman three times as "this one" (Gen 2:23). This feminine pronoun (Hebrew *zō't*) is used on its own to refer to a woman only in this group of texts, and it culminates in none other than the story of Amnon's rape of Tamar, the very story that has all the other parallels we have seen to the Dinah story. After the rape, the vile Amnon has his victim thrown out, and he will not even say her name. He says, "Send *this one* away from me!" (2 Sam 13:17). The Eden story of man's first bond with woman thus sets up the two stories of a man's degradation of a woman, and the clues of language flag the connection.

And then, at yet another level, we find that such connections bind passages through the entire course of the narrative. Consider the matter of competition and violence between brothers. The story is *about* sibling rivalry, and especially about brother killing brother, from beginning to end. After all, it starts with Cain and Abel, and it ends with a story of King Solomon executing his half brother Adonijah! But sibling rivalry is not just a bookend, or even just a persistent theme. It is a connected, developed direction of the story. Thus, there is a classic riddle in the story of Cain and Abel: the text notes that Cain kills Abel when they are in the field. What is the significance of informing us that they are in a field at the time? Even early biblical commentators searched for the meaning of this seemingly inconsequential detail. But in the story of King David's rivalrous sons, we have the story of another fratricide: Absalom has his half brother Amnon killed (for raping Tamar). In an attempt to get David to pardon Absalom, a woman known as the "wise woman of Tekoa" comes and tells a story to David. She says that one of her two sons killed the other. In the course of her account, she mentions a seemingly unrelated detail: they fought "in the field" (2 Sam 14:6). The appearance of the same extraneous detail in both stories of brother killing brother does not seem accidental, especially in light of all the other evidence indicating that these two stories belong to a single author's work. And there is further evidence of their linkage, because references to "field" occur in other sibling rivalry stories in these texts as well. In the famous story of Jacob's appropriation of Esau's birthright, Esau comes to Jacob "from the field" (Gen 25:29). Indeed, Esau is introduced as "a man who knew hunting, a man of the field" (v. 27). And in another famous story from this collection, Joseph reports a dream he has had to his brothers that so offends them that they decide to kill him a few verses later, saying, "Here comes the dream-master! And now, come on and let's kill him!" Joseph begins his report of this dream with these words: "Here we were binding sheaves *in the field*" (Gen 37:7, 19f.).[23]

Note also another thematic connection within these stories. After Jacob acquires the blessing that was meant for Esau, Esau threatens to kill him. Their mother, Rebekah, therefore sends Jacob away because, she says, "Why should I be bereaved of the two of you as well in one day?" (Gen 27:45). Readers generally understand her words about losing both of them to mean that Jacob would be killed and that Esau would then be executed for murder, leaving Rebekah childless. Again, compare the wise woman of Tekoa's story in 2 Samuel 14. She pleads with King David not to let her son be executed for the murder of his brother, and the basis of her plea is that then she will be left childless. And both of these stories come in the shadow of the Cain-and-Abel story, in which God must put a sign to protect Cain, because otherwise everyone will seek to kill him for murdering his brother (Gen 4:13–15).

The point is that the same devices of language and theme that everyone can see binding the shortest units of the story also bind the largest units. But these linkages occur in an even more pervasive form. Consider the biblical story of Jacob. It is an exquisite creation artistically, an intriguing portrait psychologically, and an interpretive treasure-house—but it has always been a problem. Even Sunday school children ask why the hero Jacob, the great patriarch, withholds food from his own brother, Esau, to get his brother's birthright, and why he lies to his blind father, Isaac, on his deathbed to get his father to give him his brother's blessing. And it is not Jacob alone who is portrayed as a deceiver. His mother, his uncle, his wife, and most of his sons become involved in deception as well.

Stories of infectious conflict within a family have long been a source of powerful literature, from Agamemnon's household to Hamlet's to the Karamazovs'. It hardly seems necessary to explain why this subject matter is so fruitful artistically and so effective. It should not surprise us, therefore, to find such a story in the Bible. The Jacob story in J is manifestly about conflict from the beginning. In a most fundamental depiction of sibling rivalry, Jacob and his twin brother, Esau, fight in their mother's womb. Esau is born first, but Jacob comes out from the womb seemingly trying to hold him back. Because Esau's birth priority is set down as a given fact, Jacob's conflict with his brother requires some means of circumventing that fact. The means is deception. The story becomes an account of a family whose members are constantly manipulating one another: Jacob manipulates Esau in order to get Esau's birthright for himself. Jacob, with his mother, Rebekah, deceives his father, Isaac, in order to get for himself the blessing that Isaac meant for Esau. Jacob's uncle Laban deceives Jacob in the matter of Jacob's marriages to Laban's daughters: he promises his daughter Rachel to Jacob but gives him her sister Leah instead. Jacob deceives Laban over ownership of their livestock. Jacob's sons deceive him over their sale of their brother, Joseph, into slavery. Joseph deceives his brothers over their fate in Egypt. Jacob's sons Simeon and Levi deceive the people of Shechem over the injury to their sister, Dinah. His son Judah deceives his daughter-in-law, Tamar, over her marriage to his son. Tamar deceives Judah in return to expose his wrong.

Go try and raise a family.

The theme of deception as reflected in these stories has not gone unnoticed. Literary and theological interpreters of Genesis have observed and commented on it for centuries, and they are troubled by it. What are we to think of Jacob's behavior—and of Rebekah's and

Judah's? Some interpreters try to vindicate Jacob, to rationalize his actions, to disparage Esau, or to come to terms in other ways with what Jacob does. But my present concern is with how the writer constructed this in the text itself. The fact is that for every act of deception there is an ironic recompense later in the narrative. And the consistency of this recompense indicates that the author was developing the theme of deception and its recompense to a denouement in the conclusion of the narrative. Thus:

Jacob's first action is to obtain Esau's birthright. Esau is famished and asks his brother for food, and Jacob insists that Esau sell him his birthright in return for some lentil stew. Esau says that his birthright will do him no good if he is dead of hunger, and he capitulates (Gen 25:29–34). Jacob's recompense comes years later in an equally famous part of the story. Jacob loves his cousin Rachel, and he agrees to work for her father, his uncle Laban, for seven years for Rachel's hand in marriage. But at the end of the seven years, on the morning after the wedding night, Jacob finds that Rachel's older sister, Leah, has been substituted for her. When he asks his uncle/father-in-law, "Why have you deceived me?" Laban answers:

> It's not done like that in our place, to give the younger one before the firstborn.
>
> (Gen 29:26)

Not "the younger one before the *older,"* but "the younger one before the *firstborn."* The man who has taken away the firstborn privilege of his brother now suffers because of the firstborn privilege of his beloved's sister!

Here, by the way, is a case in which translation makes a difference. The Targum, the Septuagint, and some recent translations (New English Bible; New Jewish Publication Society) all make it "the older," not "the firstborn," thus unfortunately missing this point and unwittingly hiding it from their readers. Several scholars have discussed this irony in various terms.[24] Now, is this reference to Leah as the firstborn a chance detail based on a coincidence of language, or is it an essential development in the structure of the story? The evidence that I observed in the rest of the narrative indicates that it is indeed an essential, designed development. There are more cases of deception and ironic recompense than have been recognized until now, and the recompense in each is tied by signals in the text to convey that it is a punishment to suit the crime.[25] Consider:

Jacob's second action is to appropriate the blessing that his father intends for Esau, a blessing of prosperity and dominion. Isaac, now old and dim-eyed, not knowing when he will die, asks Esau to hunt some game and prepare a meal for him, after which he will bless him. Rebekah prepares a similar meal and sends her preferred son, Jacob, to pose as Esau to his dim-eyed father and thus to get the blessing. Jacob wears his brother's clothing and the hides of goats to imitate his brother's smell and hairy skin, and Isaac gives him the blessing meant for Esau. When the deception becomes known, Isaac trembles, and Esau cries in anguish, "Was his name really called Jacob [meaning 'to catch, usurp']! And he's usurped me two times now!" (Gen 27:36).

Years later, Jacob is paid back for this as well. When his own sons, jealous of his preferred son Joseph, sell Joseph into slavery in Egypt, they dip Joseph's coat of many colors in the blood of a goat, and they say to their father, "Recognize this." He says, "My son's coat. A wild animal ate him. Joseph is torn up!" (37:32–33). Jacob had deceived his father with his *brother's clothing* and the meat and hide of a *goat*. Now his sons deceive him with their *brother's clothing* and the blood of a *goat*. (Worse, really he deceives himself in response to their request that he "recognize.") And the word used for "goat" in the second story is the Hebrew *śāʿîr;* and Seʿir is the name of the place where *Esau* settles (33:16). And Esau is described as a "hairy man" (Hebrew *śāʿîr*), which is the reason for the goatskin deception in the first place (27:11). Is the phrase "punishment to *suit* the crime" not particularly appropriate here?

Laban, too, suffers recompense for deceiving Jacob over Rachel and Leah. Years later, Jacob deceives Laban back. He agrees to work for Laban in return for all the spotted, speckled, and brown sheep and goats among Laban's flocks, and he then connives (by magic? miracle? paragenetics?) to produce such colored sheep and goats in great numbers (30:28–43).

Like father, like sons: Jacob's sons, in their own sibling rivalry, had deceived their father over the sale of Joseph into Egyptian slavery. Years later, when Joseph has risen to extraordinary power in Egypt, he deceives his brothers. Jacob sends the brothers to Egypt to buy food during a famine. When they arrive in Egypt, Joseph recognizes them, but they do not recognize that this powerful Egyptian is their brother whom they sold when he was seventeen. Repeatedly they refer to themselves as "your servants" as they speak to the brother whom they once sold as a servant (42:10–13; 44:9, 16, 23, 31). Again, the deception-for-deception theme is developed subtly in the text. To review: ten brothers had sold Joseph into slavery for twenty measures of silver. Years later, when they come to Egypt, Joseph accuses them of being spies. They deny it. They are, they say, ten of twelve sons of a man from Canaan, one son having died (Joseph!) and the other (the youngest, Benjamin) having remained home with his father. Supposedly to test whether they are telling the truth, Joseph keeps one brother as a hostage and sends the other nine back to Canaan with instructions to bring Benjamin back with them. On the way, they discover that the silver that they have paid for the food has mysteriously been put back in their *(nine)* sacks. Later, they return with their youngest brother, Benjamin, to buy more food. Joseph then sends off all eleven brothers; but, again, he secretly has their silver placed back in their *(eleven)* sacks. The total number of portions of silver returned is thus *twenty,* which happens to be the same as the price for which the brothers had sold Joseph (37:28; 42:19; 43:15–23a, 24–34; 44:1)! The brothers thus receive twenty measures of silver a second time, but this time under very different circumstances. They learn what it means to feel helpless, to be the victims of injustice.

Like father, like son: Jacob's son Judah deceives his daughter-in-law, Tamar. Two of Judah's three sons have been married to Tamar, and both have died. He promises her that she may marry his third son, Shelah, but he says that Shelah is still too young for marriage and that she should wait a while. She has a long wait. Shelah grows up, and Judah makes no wedding plans; he is unwilling to risk losing his last son over her. But Tamar knows something of the art of deception herself. She poses as a veiled prostitute and attracts Judah's attention. He lies with her; and he leaves articles he was wearing (seal and cord) with her as a pledge that he will pay her one goat. Some time later, Judah hears that his daughter-in-law is pregnant. Since she is still officially bound to his son Shelah, this is the equivalent of adultery. Judah says that she should be burned, but she produces his own pledge items and says, "Recognize these." He does, and he acknowledges that he has wronged her. Again the deceiver has been deceived, and again the recompense fits the original deception: he denies her a man; she makes him her man. *And* the deception involves attire and a goat. *And* it culminates with a "Recognize." *And* she gives birth to *twins*. And they struggle to be first out of the womb (Genesis 38)!

Three other sons of Jacob commit acts of deception that prove costly. His firstborn son, Reuben, sleeps with Jacob's concubine Bilhah, and Jacob finds out (35:22). His second-born and third-born sons, Simeon and Levi, deceive the people of the city of Shechem. I have already made several references to that story, in which the prince of that city has had sex with their sister, Dinah, and only later asked for her hand in marriage. Her brothers respond to the proposal with deception, saying that they will permit intermarriage with the Shechemites only if the latter circumcise all their males. The Shechemites agree to the surgery, and, when the Shechemite men are immobile from the pain of the circumcisions, Simeon and Levi come and massacre the city. Jacob criticizes them for this act, but they answer, "Shall he treat our sister like a prostitute?" (Genesis 34). The price that Reuben, Simeon, and Levi pay for their deceptions is the loss of their places in the succession to the birthright. On his deathbed, Jacob explicitly takes away Reuben's preeminence, and he condemns Simeon and Levi to dispersal (49:3–7). Every deception is paid back.

Now, the same theme of deception-for-deception that we see here in J resurfaces as a linkage within the story of David and his royal family in the Court History, and the same devices are used to flag the connections for the reader. As with Jacob, so the story in the Court History begins with its hero, David, committing a hurtful and deceptive act, which brings a chain of ironic recompense in the family. David has an affair with a married woman, Bathsheba, she becomes pregnant, and David arranges the death of her husband, Uriah, and marries her himself. A prophet, Nathan, tells David that the consequences of his actions are:

1. There will be continual violence in his house.
2. Another man will take David's wives publicly.

And so it happens. David's son Amnon rapes his half sister, Tamar. Tamar's brother Absalom has Amnon murdered. Absalom later claims David's throne. There is war between David and Absalom. David's general, Joab, kills Absalom. Next, David's son Adonijah claims David's throne. David appoints Bathsheba's son, Solomon, instead. Solomon later executes Adonijah—and Joab as well. David's house is racked with violence.

As for the prophecy that a man will take David's wives publicly, the man turns out to be his own son, Absalom. As a part of his claim to his father's dominion, Absalom makes a public display of having sexual intercourse with ten of David's concubines. And, as in the story of Jacob in J, there are signals in the text that point out the ironic recompenses. In the Court History, David sees Uriah's wife, Bathsheba, bathing *from his roof,* which first attracts him to the affair. Later, the way that Absalom publicizes his displacement of David is:

> And they pitched a tent for Absalom *on the roof,* and Absalom came to his father's concubines before all Israel's eyes.
>
> (2 Sam 16:22)

And later still, the way that David learns of Absalom's death is that a watchman *on a roof* sees the runners coming with the news.

This now gives meaning to all the parallels I mentioned at the beginning between the J story of Dinah and Shechem and the Court History story of Amnon and Tamar. All of the irony that pervades the story of David's sin and recompense is doubled when we see it to be itself an ironic recurrence? reenactment? variation? legacy? of what happened to David's ancestor, Jacob.[26] Similarly, Tamar's torn coat of many colors is a sad reminder of the other torn coat of many colors that also was worn by a young, innocent victim of violence at the hands of brothers, Joseph. And these sequences of recompense in J and the Court History are connected: The J story of Jacob's deathbed blessings results in his eliminating the first-, second-, and third-born sons, resulting in the preeminence of *Judah,* which is David's tribe. And the J story of Judah and Tamar culminates in the birth of their son Perez, whose descendants are destined to become the major clan of Judah—which is fulfilled in David, who was understood to be from Perez (Ruth 4:18; 1 Chr 2:4–15).

And so on with all of the connections I have mentioned between these two narrative sequences. The chains of words and phrases that occur only in these texts were evidence that *something* was going on here. The actual contexts in which these words and phrases appear now show us what that something was. An author has developed a tale of how the acts of the parents, including their errors, reverberate in the lives and acts of their children and even of their descendants many generations later. Now we can appreciate the famous self-description that this author ascribes to the deity in the ultimate revelation, the moment in which Moses actually sees God at Sinai. God is described as:

> keeping kindness for thousands, bearing crime and offense and sin; though not making one innocent: *reckoning fathers' crime on children and on children's children, on third generations and on fourth generations.*
>
> (Exod 34:7)

This is a book of generations, of continuity, of recurrence and recompense.

Indeed, perhaps we should look for the root and model of this human inclination for deception in the narrative even prior to Jacob. After all, Jacob's father, Isaac, tries to deceive the Philistine king Abimelek into believing that Isaac's wife, Rebekah, is Isaac's sister (Genesis 26). Isaac's father, Abraham, had tried the same deception on the Egyptian king (Genesis 12). Does the chain of deceptions begin with Rebekah, who helps her preferred son, Jacob, to deceive Isaac? Or does it begin with Isaac? Or with Abraham? Or with the snake?! The deception theme explicitly starts in Eden: When God questions Eve, she says, "The snake tricked me." And the theme pervades this narrative's episodes. When God questions Adam, he tries to shift the blame to Eve and even to God; he says, "The *woman,* whom *you* placed with me, *she* gave me from the tree, and I ate." When God questions Cain about the whereabouts of his missing bother, Abel, Cain says, "I don't know. Am I my brother's watchman?" The rebels Dathan and Abiram accuse Moses of deceiving the Israelites in the wilderness; they say,

"Will you put out those people's eyes?" (Num 16:14). The prostitute Rahab deceives the king of Jericho when she helps two Israelite spies to escape. Some Gibeonites trick Joshua into signing a pact that prevents the Israelites from attacking them (Joshua 22). Several of the battle stories in the books of Joshua and Judges involve ambushes that delude armies and lead to their demise. In the stories of Samson, both his wife and later his lover, Delilah, are told, "Trick him" (Judg 14:15; 16:5). David's wife Michal saves him from her father, Saul, by putting idols in David's bed and a goat-hair pillow (as in J, where there's a goat, there's a deception) to deceive her father's men about David's escape (1 Sam 19:13). And Saul says to her, "Why have you deceived me?" (19:17), just as Jacob had once said to Laban, just as a woman at En-dor later says to Saul himself (1 Sam 28:12).[27]

Do we get deception elsewhere in biblical narrative as well? Certainly. But not nearly in this concentration. Elsewhere it is an occasional element in stories. Here we have seen that it is a theme that binds. The deception construction is one prominent element helping to unite the narrative into a continuing, coherent whole. The element of fratricide likewise contributes to this coherence of the work: there are at least seven stories involving a brother killing his brother. And a variety of other such threads run through the text, bind it, and shape it into a meaningful story.

Compare, for example, the language and technique of Adam's wriggling to get out of it when he has been caught red-handed to King Saul's wriggling when the prophet Samuel catches him keeping some booty he was commanded to destroy (1 Sam 15:13–26). Even when Saul admits his sin, he tries to mitigate it by saying that his error was giving in to the *people's* wishes. He says, "I listened to their voice." When we recall that Adam tried to mitigate his actions by blaming it on "the *woman,*" and that the deity rejected this argument by noting that "you listened to her voice," we can predict just how far Saul is going to get with this argument even before we finish reading the episode.

But, even more, when you read it as a continuous work, there is a chance that when you read about Saul you will be *reminded* of Adam—as I was. And then you may find it all the more fitting that the scene with Saul ends with his accidentally tearing Samuel's clothing, which is the mark of his loss of the kingdom because of his sin. It is an ironic twist on the fact that clothing was invented by the deity in the end of the scene with Adam.

Similarly, the story of the people who surround the house in Benjamin becomes not an uncreative shadow of the story of the people who surround the house in Sodom. It is rather a story of a recurrence of tremendously dangerous events, with a new denouement: The first time, at Sodom, God personally deals with the perverse city. The second time, in Benjamin, the human community itself reacts to the city's crimes.

The same can be said of the Dinah-and-Shechem and Tamar-and-Amnon episodes, and of Cain-and-Abel and Solomon-and-Adonijah, and of so many more. The feel for relationships in the biblical text that was lost when biblical source criticism started is restored when we recognize the unity of this lengthy work. And when it is read without the intervening texts from other sources, the relationships are more obvious, more immediate, and often more powerfully felt.

That is the human side of the story. But, at the same time, the story never ceases to be about God. The author of J has been described as "the Bible's first theologian," but that description uses the term "theologian" extremely figuratively. The author may be conveying a particular theology, yet does not do so systematically but rather by telling a story involving divine-human relations. That story deals with the acts of God, but only those acts that relate to human beings. It deals with the words of God but, with a very few exceptions, only with words that are spoken to human beings. It does not deal at all with the nature of God: what is God? where is God? Like all of the other writers of the Hebrew Bible, the writer of this work does not picture, define, identify, or locate the deity.

What the work does provide is a description of the personality or character of God. The one story that describes the divine character outright is in Exodus 34, the scene of the ultimate human experience of God, and in that scene God is described as torn between justice and mercy. God's self-description there states on the one hand that the deity does "not make one innocent: reckoning fathers' crime on children and on children's children, on third generations and on fourth generations." But, at the same time it states that the deity is "merciful and gracious, slow to anger and abounding in kindness and faithfulness, keeping kindness for thousands, bearing crime and offense and sin." This conception of a God who is torn between justice and mercy informs the whole work. This is not the "Old Testament God of wrath" that people frequently (and erroneously) imagine. It is more a picture of a God who can get angry but cannot stay angry.

Thus: The deity sees the quantity of human wrongdoing, regrets having made humankind, and decides to "wipe out the human whom I've created from the face of the ground." But God does not wipe them out. One human, Noah, and his family are spared, and the species is preserved even though the deity explicitly recognizes its inherent inclination to do wrong (Gen 6:5; 8:21).

Thus: When Moses' spies bring back a partly negative report of the promised land, the people are fearful and want to return to Egypt. The deity states a plan to destroy them and start a new people descended from Moses, but Moses quotes God's own words back: "slow to anger and abounding in kindness, bearing crime and offense . . ." (Num 14:18). And God forgives.

And so on repeatedly. Heroes suffer for their offenses but are ultimately successful. The people of Israel suffer for their breaches of the covenant that God has made with them, but they are saved from one crisis after another. I do not mean to ignore the violence at divine command that this writer depicts, especially in six chapters in the book of Joshua that describe battles between the Israelites and the Canaanites (6–8; 10–12). War, violence, and suffering were realities in this author's world, and the author does not exclude them from the story. But this author does not present the uniformly angry, harsh picture of God that some writers have imagined. Much to the contrary, by developing both sides of the divine character—angry and loving, just and merciful, punishing and forgiving—the first great prose writer set the theology that informed all the works of the Bible that were later added to this story.

The author also varied the depictions of the way in which the deity acts in the world. In some cases, divine presence is manifest through miracles, angels, and direct speech to individuals. In other cases, divine activity takes place behind the scenes. Thus God walks in the garden in Eden, talks to the humans, personally makes their first clothes, personally closes the ark, and appears to Abraham. Abraham, Sarah, Hagar, and Lot encounter angels. But the hand of God is more subtle in the stories of Jacob and Joseph. The chain of recompense for deceptions suggests to the reader that something is going on, but it is not accomplished by blatant miracles and revelations. The divine actions become manifest again in the stories of the drying of the Red Sea and the Jordan River and the other miracles of the stories of Joshua. But they are subtle again in the stories of David, as the seemingly mundane struggles within the royal family come out fulfilling a prophecy. Similarly, a prophecy to Rebekah about the twin babies, Jacob and Esau, who are in her womb and a prophecy to Samson's mother while Samson is still in her womb affect the destinies of their future offspring.

In this work, God is sometimes manifest and sometimes hidden, sometimes just and sometimes merciful. One might be inclined to wonder whether these differences reflect multiple authors after all. But that is not the case. The different aspects of God fall within the texts that share the common terms and themes, the texts that connect to one another and allude to one another. The complexity of the divine in these works is not the result of mixing texts by two or three authors. It is the result of one author's rich conception of God. This should not come as a surprise when we consider how rich is this author's conception of human beings and of human relations.

One thing that is not dual about God in this work is gender. Like other biblical authors, this writer lived in a community that pictured the deity as male, and this writer in turn depicted God in masculine terms. The divine name, YHWH, is as masculine a name as Abraham or Isaac. (The feminine would be THWH.) References to the deity come with masculine pronouns, verb forms, and adjectives. Whatever our present conceptions of God, whatever our present sensitivities to equality of the sexes, the fact remains that in the first stage of monotheistic religion God was conceived of as masculine, and this work comes from that era.[28]

This author constructed a tapestry of interactions between this God and the humans whom He fashions. This is the story that the first great prose writer's work tells:

> YHWH makes a world that begins as a paradise, but paradise does not work out because conflict and deception are apparently inherent. Humans get the divine ability to make judgments of what is good and what is bad. As they incline toward bad, YHWH starts over with a virtuous man, Noah, but this does not solve the problem, because a few generations later humankind

has gone bad again. YHWH covenants with the family of a virtuous man, Abraham, with the explicit intent that this relationship will ultimately bring blessing to all the families of the world. The narrative then follows this family's destiny through twelve generations, from their internal conflicts to their growth into a nation. They are enslaved, then freed. Their rebellious unpreparedness for freedom leaves YHWH considering the possibility of starting over with one man once again: Moses. But Moses appeals to YHWH's compassion, and the nation survives to return to the land that YHWH promised to Abraham. There follows an age of sporadic wars. Lacking a king, each person does "what is good in his eyes," and the nation comes into conflict with other groups (Ammonites, Philistines, etc.) and among themselves. Kings are established—first Saul, then David—and the focus becomes, as at the beginning, the destiny of a family, especially their conflicts; but this time their family struggles have blatant consequences for the nation. The conflict reaches its climax and conclusion as the throne passes securely from father to son—from David to Solomon—and Solomon eliminates his main potential threats, including his half brother, and establishes a secure kingdom sanctioned by a divine covenant.

That is an outline of the story that this writer tells, from the establishment of the world to the establishment of a kingdom. At present, there are many popular novels that tell the story of a family through several generations. This work is extraordinary in that it narrates such a saga through twelve generations (a biblical number!) from Abraham to Solomon. It is a masterpiece of continuity. It connects the actions and fate of each generation with those that come before and after it. And it sets this saga in the context of all of humankind by preceding it with a short section containing stories of creation, Eden, the flood, and the tower of Babylon. And it links the family saga to that worldwide concern by saying in the first words of the deity to Abraham that his descendants are to become "a blessing to all the families of the earth."

About the Author

The basic description of this author that I gave in *Who Wrote the Bible?* still holds true. The author lived in the kingdom of Judah, most probably in the latter ninth century B.C.E. (later than the time of the independence of Edom from Judah), but certainly before the Assyrians' destruction of the kingdom of Israel in 722 B.C.E. The author was from a class that was educated and had access to writing materials. The author was a writer, not a collector of oral tales. The work does not have the character of oral composition, though it is possible that the author incorporated elements from oral sources. It has recurrent themes, but it lacks the kind of formulaic repetition that is typical of oral poetic tales. The author was probably a layperson, not a priest. The work is different from the biblical works that are ascribed to authors who were priests (including E, P, and D) in many ways; notably, unlike the other source works of the Five Books of Moses, it does not contain a lengthy code of law. It is rather the work of a literary artist, whose aim seems to have been partly that of the writer and partly that of the historian—to tell the story, and to tell it beautifully.

Also as I suggested in *Who Wrote the Bible?* this author, unlike the authors of those other source works, may have been a woman. This suggestion must be taken with great caution because it is a subtle task at best to determine the sex of an author of a three-thousand-year-old work by clues within the text. That text clearly conceives this to be a man's world. After all, it begins with a pronouncement on woman in Eden: "Your desire will be for your man, and he'll dominate you." Yet it still depicts women achieving a measure of power within this male-led structure through strength of character (Abigail), cleverness (Rachel), influence on sons (Rebekah), pestering (Delilah), deception

(Potiphar's wife), and negotiation (Rahab). And it is a woman, Eve, who first pronounces the name of God: YHWH (Gen 4:1). Because of the complexity of this point, I still am not prepared to state with real confidence that the author was or was not a woman. I raised the matter of the sex of the author of J in my earlier book simply because I found it strange that scholars had considered the geographical location of authors, the century in which they had lived, their social class, their standing as priest or layperson, and their opinion of monarchy, but not so basic a matter as whether they were men or women. The literary critic Harold Bloom took up my suggestion and transformed it into a full-blown claim that the author of J was in fact a woman. Ironically, in light of what I am now saying about this writer's full work, Bloom spoke of the author of the Court History as her male rival! One will understand my lack of trust in Bloom's confidence in his sense that the author of J was a woman, given that he sensed another part of the same author's work to be by a man.

Since I've mentioned Bloom, I should say that I was disappointed at his claim of originality about the author's sex in *The Book of J,* but I'll also pay him his due: He had a sense of authors, of the living, breathing persons who write the books that we read, including the Bible. The wave of recent biblical criticism that insists on focusing on "the text as it is," somehow separated from the persons who fashioned it and from the world in which it was born, has a certain value; but the success of Bloom's book shows that ultimately such criticism must fail to provide something that readers obviously crave. It should not have taken a book like Bloom's to demonstrate this point, but since it did happen this way, we might do well at least to take note of the point. The separation of the study of the Bible's qualities from its authors, which also means the separation of literary study from history, in the end is just not sufficiently satisfying.

Similarly, some scholars speak of the last editor of the text as if that person were the one responsible for the Bible in the form that we now have it: as if that person had conceived of the puns, the ironies, and the persons and events of the story. But this is not true. In the first place, it is an incorrect picture of what happened, a picture that takes away the credit from the writers who actually created these things. And second, there was no such single editor. There were several. And editors were as different from one another in biblical times as they are now. Some editors appear to have tried to retain as much as possible of their source works. Such is the case of the last editor of the Torah, who is known in biblical scholarship as the Redactor (for short: R). And such is the case of the editor who assembled the books from Deuteronomy through Kings, who is known as the Deuteronomistic historian (for short: Dtr). Frustratingly for us now, one of the earlier editors was more willing to make cuts in his source works. The editor who combined the work of the first great writer with another work, the text known as E, in the books of Genesis, Exodus, and Numbers (known as RJE) removed some portions of each. As a result, there are a few gaps in the first part of the story. In my translation of the work here, I have marked the gaps with brackets. Fortunately, it appears that the vast majority of the work remains intact.

For most readers, the first portion of the work will be familiar, containing many of the most famous stories of Western literature: Adam and Eve, Abraham, Moses. As they read on, however, they will encounter less well known tales, and there will be surprises and discoveries in store. And, even for more experienced readers of the Bible, there should be a plethora of new finds, connections, and meanings to be met and enjoyed. Here are some of the things that you are likely to notice or feel:

The flood story is half as long as the way you have read it in the past. So is the story of Joseph, and the story of the spies, and the story of King Saul and the young David. There is no law except the Ten

Commandments, and they are different from the ones that are usually quoted. Family matters are present in a much higher proportion than usual. Sexual matters are present in a much higher proportion than usual. The lives of individuals and families are linked to major events, especially wars. The middle portion of the story deals with an age of war, and the vivid battle descriptions may be jarring—but, happily, the work ends in an age of peace—under King Solomon, whose name means "peace." There is also the matter of what you will *not* find here. Some readers may be struck by the absence of famous biblical stories that were not part of this work, including the seven-day creation, Abraham's near-sacrifice of his son, Isaac, the golden calf, and the judge Deborah. They are wonderful stories, but they were by other authors, and they do not fit the language and structure of this work. Here, above all, you are likely to feel a coherence in the full story to a degree that you have never experienced before.

And so I present this text not as a collection of folktales preserved from unknown oral storytellers but rather as the crafted design of a gifted writer. It is preceded by a message titled About the Translation to explain my standards and approach to translation. Because it is best for me to make parts of the case for the unity of this work only after you are familiar with its full narrative, I have included an Afterword following the translation. There I shall provide some additional (and, to my mind, even stronger) literary evidence for its being a single work, and I shall offer some final comments. Textual Notes contains some notes on individual difficult verses and words. There is also an Appendix, in which I list and analyze the details of the case for its unity and high antiquity. That presentation is intended for scholars, but I have tried to write it in such a way that it will also be accessible to any interested laypersons who have sufficient background in this subject to look into it.

There is the matter of the book's title. I have agonized over choosing a title that would be appropriate and worthy of it. In the end, though, I do not believe that I or anyone else has the right to title it. That is the author's prerogative, and we are the readers, not the authors. The custom in the ancient Near East was to call books by their opening words. Thus the book of Genesis has always been known in Hebrew as *bᵉrēšît,* "In the Beginning." I therefore understand this work to be titled by its own first words, "In the Day."

In the Day is a treasure, a gift to us as the turn of the millennium is approaching. It is filled with lessons about the threat of conflict within families and between nations, about the power of love and the consequences of betrayal, about human beliefs and attitudes toward God, about relations between men and women, parents and children, siblings and siblings. It comes to teach us, in our generation in which all values are denied, that good and bad have meaning.

I am writing this in an age in which that meaning seems to have been lost. Students learn the little formula "What I say is good, you say is bad." All values are relative—especially moral values—but also standards of art, of what is good literature, and of what is true. This is not some deep concept—that we have transcended old-fashioned ideas of right and wrong. It is more like little children who have found an intellectual key to a forbidden cookie jar. It is not Nietzsche's concept of being "Beyond Good and Evil." Our time might more properly be described as *beneath* good and evil. If this were just an intellectual game that we played in school and college, it would be one thing. But it is not harmless. In our confused century, we have seen suffering beyond counting, comparing, or describing.

And as this age of sailing without a compass goes on getting less and less secure and more and more frightening, suddenly this work comes back from nearly three thousand years ago. And it says that, yes, humans have the power to make judgments of what is good and bad and right and wrong. In this story, the creator of the earth does not always *reveal* what is good and bad, but rather the humans take the fruit that enables *them* to make these judgments. There is no claim that humans have insight to know "absolute" good or bad. This is not a denial that our judgments are relative, and it is not about debates over acts that are considered good in one culture and bad in another. It is a recognition that humans make choices and that the choices have consequences. The *knowledge* of good and bad means learning what each produces.

This work conveys that deception and cruelty in a family fester and spread, and that the most likely way to stop it all is for one member of the family who is entitled to recompense to choose not to take it—and forgive. It implies that human inclinations that lead to war—which is the worst thing in the world—must give way as we see war's results. And it teaches us that the choices and actions we make have implications through a vast span of generations of our descendants. Whether one believes that we acquired this ability to make judgments of good and bad by that single act of eating a fruit or that we acquired it through centuries of experience, the fact is that we have this power. And this work comes to tell us that this power has effects, and we must use it with wisdom. Its discovery at this millennial juncture in human history feels to me almost like a miracle, a gift to us from our ancient past. Now I invite you to read it and observe for yourself: its unity, its coherence, its power, its technique, its psychological insight. Experience, after nearly three millennia, the great work that became the core of the Bible and the beginning of prose literature.

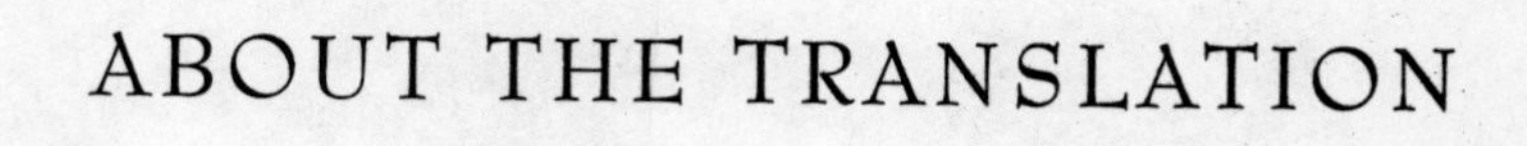

ABOUT THE TRANSLATION

EVERY NEW TRANSLATION OF THE BIBLE BRINGS ITS OWN SET OF PURPOSES IN ADDITION TO THE TRANSLATORS' PARTICULAR SKILLS AND FEELINGS FOR THE TEXTS. Some seek the most literal possible rendering of the text. Some seek to produce the most felicitous possible reading in English. Some seek to reproduce the feeling of the Hebrew: to capture its puns, alliteration, and rhythms. Most translators seek some combination of these things, and they differ in the particular extent to which they favor one purpose or another. My purpose in this translation is somewhat different from these. Since I am presenting a particular work for the first time, it is especially important that I convey the text of the original as carefully as possible. So, for example, when a wording is unclear in the original in a few cases, I do not try to clear it up in the translation. I use a translation that retains the unclarity of the Hebrew. This may be annoying or frustrating to readers. But that is the point: it is annoying and frustrating as one reads it in the original. My task is to translate the Bible, not to make a better Bible. I have to accept the fact that I am the translator, not the author. This does not mean that I have given up on producing a felicitous English. It means that, if the reader finds this work to be a "good read," as I hope you will, it is primarily because of the qualities of the original, and not because of my own refashioning of its language. And when you find an occasional awkwardness, I hope that you will find it worthwhile to join the biblical scholars in encountering the difficulties of the Bible's text rather than having the scholar try to solve it for you or hide it from you. In the end, I hope that you will agree that, even in translation, the power and beauty and sensitivity of this work shine through.

My expectation is that most of my colleagues in the field will want to note the verses as listed on p. 12 and read the text in the Hebrew. This translation is primarily for persons who are not sufficiently advanced in Hebrew to read the text in the original. Therefore, this is not an annotated critical edition of the text. Because I want readers to be able to feel the continuity of this work, I did not want readers' attention to be interrupted by footnotes. Someday an annotated version should be done, with full text-critical apparatus, but the purpose of this first translation of the work is to provide an opportunity to read the work from beginning to end without interruption. I have provided Textual Notes on some specific points, on pp. 305–324. I have translated the Masoretic Text (MT) generally, with only a few changes based on Greek or Qumran (Dead Sea Scroll) versions, none of which is crucial to the points that I am emphasizing elsewhere in the book. (Citations of chapter and verse refer to the Masoretic Text. In some English translations there are occasionally differences of a verse or two, or even several verses, in their numbering. For example, Gen 32:4–13 in the Hebrew text is Gen 32:3–12 in many translations. Readers whose practice is to look up cited passages when reading books about the Bible, or who want to compare the translation here with other translations, are hereby advised to take this into account.)

Following are some notes on a few basic points of this translation:

1. The original uses very simple language: "big," "very much," "heavy," "bad." When translators turn it into more sophisticated language and enrich the vocabulary by using multiple English translations for the same Hebrew word (for example, "big," "great," "mighty," etc., for *gādōl*), they make it sound better in English but mask the simplicity and consistency of the original. I am especially moved to keep the multiple translations of single Hebrew words to a minimum, precisely to let readers become aware of this quality of the language and of this author's way of making so much out of a relatively small bank of vocabulary. This should also help readers to get a feeling for the consistency of language through this work.

   So, as a major example, the Hebrew word *rā'* has a tremendous range: from "bad" to "harm" to "terrible" to "evil" to "wretchedness." But I translate it in almost every occurrence as "bad," even though this will seem like an insufficiently strong word in several places, in order to convey the important point that this work's story begins with the eating from the tree of knowledge of good and *bad,* and that this introduction of human knowledge of bad begins a chain of actions that will inform the story until its end.

Similarly, I translate Hebrew *har* consistently as "mountain," though it can refer to a large mountain or a small hill.

This goes for the Hebrew conjunction as well, which begins almost every verse. It usually means "and," but it has a wider range of meanings as well, so translators make it "but," "since," "while," and more. I leave it as "and" in English except in cases in which the text absolutely forces us to take it differently. Further, some recent translators simply leave it out altogether; so, unlike the KJV, RSV, and JPS, each sentence in these translations does not begin with the word "and." On this point, too, I prefer to retain the feel of the original. I retain the word "and" where it occurs in the Hebrew.

2. Many translators eliminate old English terms—the "whither's" and "thither's" and "whence's" and "thence's" and "hence's" and "thee's" and "thy's" and "thou's"—to produce a contemporary translation; yet they still retain some old terms that do not have ready counterparts in contemporary English, such as "lest" and "in the midst." The result is unfortunately an English that no English speaker ever wrote or spoke. And so it just doesn't feel right. I have tried to produce a translation that is consistent in the English it employs. Sometimes there is simply no way to convey a Hebrew phrase's meaning in contemporary English, but I have tried for this consistency to the extent that it is possible while being true to the original.

3. Sometimes I keep an idiom rather than translate it away—so long as its meaning is clear—because then my translation makes known to the reader of the translation how the Hebrew works. For example, in Gen 16:2 I translate the Hebrew quite literally as: And Abraham "listened to Sarah's voice," which is unusual and even a bit redundant in English, but I would rather have the reader of the translation see how this is expressed in the Bible than modify it to something like "listened to Sarah." Similarly in Gen 19:8 and many other places, I translate the expression "good in your eyes" literally because it is understandable in English even if it is not the common idiom.
4. It is now impossible to know for certain where the paragraphs began in the original. Like all translators, I have made decisions where to begin new paragraphs based on the content, logic, and emphases of the stories.
5. Often a new translation is meant to show how new and different one can be from the old, familiar translations. The purpose of my translation, however, is to show something that has been present all along but not known. So I have not gone out of my way to be new for new's sake.
6. The name of God: I make it the same as it appears in the original, with the four consonants showing, thus: YHWH. This is known as the Tetragrammaton. In biblical times people would have read the name itself. In the period following the completion of the Hebrew Bible, Jews began the practice of not saying the divine name out loud. Christians followed this practice as well. The practice for centuries was to say "the LORD" (Hebrew *'ădōnāy*) whenever one came to this word rather than to say what the four Hebrew letters actually spelled. Currently some people have returned to the practice of saying the name out loud. Readers should follow the same

practice in reading this translation that they would follow when reading the original. If they do not pronounce the divine name aloud, then they should say "the LORD" whenever they see the four letters.

In the first chapter of this work, unlike the rest of the Torah, the divine name appears together with the word "God" *('ĕlōhîm)*. The addition of the word "God" appears to have been made by the Redactor at the time that this work was combined with P so as to make the transition from the Priestly creation story, which uses "God" alone, to the J story, which uses the divine name alone in narration. These additional words thus did not appear in the original work and therefore are not included in the translation here. (The divine name appears together with *'ĕlōhîm* rarely in the narrative even outside the Torah, as in the exceptional 2 Sam 7:25.)

7. The gender of God: Even though, like most people, I do not conceive of the deity as male or female, there is no way around the fact that the author of this work did in fact present God in consistently masculine terms. Even the name of God is masculine. (The feminine would be THWH.) I have therefore conveyed the masculine Hebrew conception in the translation as well. Again, my point is that I am translating an original work that someone else wrote, and I do not seek to impose my theological conceptions on that person's work, nor do I want to hide that person's views by means of a translator's power.
8. I have left the book, chapter, and verse numbers in the text so that readers can compare this to any other Bible translation.
9. In the portion of the work that is now contained in the Deuteronomistic history (the books of Deuteronomy through Kings), I exclude as little as possible of passages that might conceivably have been written by the editor (the

Deuteronomistic historian) rather than by the writer of *In the Day*. Even in the case of phrases that are commonly regarded as coming from the Deuteronomist's pen, we have to consider the possibility that the Deuteronomist derived them originally from this source and then chose to use them elsewhere.

10. Hebrew sometimes has the infinitive of a verb placed before the verb itself in order to convey emphasis. Thus the Hebrew *môt yûmāt* means literally: "dying he shall be put to death." Most English translators use a formulation such as "he shall surely be put to death" to convey it. Since the function of this infinitival formulation in Hebrew is to emphasize, I think it is best translated by the usual mechanisms of emphasis in English. The usual ways to convey emphasis in English are either the use of italics or exclamation points. I therefore generally use italics to convey Hebrew infinitival emphatics; occasionally I use an exclamation point. In cases where neither of these English conventions properly conveys the Hebrew meaning, I leave the infinitive untranslated.
11. Cognate accusatives (for example, "I dreamed a dream," "I did a deed") are fairly common and a hallmark of literary style in Biblical Hebrew. Cognate accusatives are not incorrect grammatically in English but are sufficiently rare as to be felt by English readers to be uncomfortable. And so, like nearly all translators, I convey cognate accusatives without repeating the cognate forms. Thus, for example, I translate *ḥălôm ḥālamtî* as "I had a dream" rather than the literal "I dreamed a dream."
12. Like nearly all translators, I convey the cohortative as "let me" or "let us" or "may I" or "may we." Unfortunately, these English forms suggest permission, as if the person using the cohortative is asking to be allowed to do something. Readers should note that the Hebrew does not usually have

such a connotation. It merely expresses the person's wish or intent, without necessarily implying that any action by anyone else is necessary.

13. The same applies to the jussive, which I convey as "let it" or "let him" or "let them" or "let there be" or "may he" or "may they."
14. The formulation "And he said, saying . . ." occurs fairly often in the Hebrew text. Though it feels redundant and awkward in English, I preferred to retain the extra word—"saying"—to reflect the original.
15. Etymologies of names. Often a story is told that explains the origin of a name of a person or place. For example, the name Isaac, Hebrew *yiṣḥāq,* means "he laughs" or "let him laugh," and it is derived in this story from the fact that Sarah laughs at the idea that she could give birth to a son in her old age (Gen 18:12–15). It is difficult to convey this in an English translation without adding some bracketed or footnoted comment. Similarly with the names Noah, Babel, Jacob, Moses, and so on, it is usually impossible to convey such Hebrew etymological material in translation. A note or bracketed insertion interrupts the story for readers, and I did not want to do that. So, as most translators have done in the past, I have left the text itself alone, accepting the fact that this is one of the inherent limitations of translation, and I explain the etymologies briefly in the Textual Notes.

[illegible]

[illegible] conditions. If there is present [illegible] the names [illegible] in [illegible] that any [illegible] necessary.

[illegible] A [illegible] of [illegible]

The [illegible] thirty [illegible] and [illegible] the extra word [illegible]

(3) Incomplete phrases. Often a story [illegible] explains the [illegible] or "let him laugh," [illegible] with the names [illegible] and [illegible]

[illegible]

ביום

In the Day

Genesis 2:4b. **In the day that YHWH made earth and skies**

— 5. when all produce of the field had not yet been in the earth, and all vegetation of the field had not yet grown, for YHWH had not rained on the earth, and there had been no human to work the ground, 6. and a river had come up from the earth and watered the whole face of the ground—

7. YHWH fashioned a human, dust from the ground, and blew into his nostrils the breath of life, and the human became a living being.

8. And YHWH planted a garden in Eden at the east, and he set the human whom He had fashioned there. 9. And YHWH caused every tree that was pleasant to the sight and good for food to grow, and the tree of life within the garden, and the tree of knowledge of good and bad.

10. And a river had gone out from Eden to water the garden, and it was disbursed from there and became four heads. 11. The name of one was Pishon; that is the one that circles all the land of Havilah where there is gold. 12. And that land's gold is good; bdellium and onyx stone are there. 13. And the name of the second river is Gihon; that is the one that circles all the land of Cush. 14. And the name of the third river is Tigris; that is the one that goes east of Assyria. And the fourth river: that is Euphrates.

15. And YHWH took the human and put him in the garden of Eden to work it and to watch over it. 16. And YHWH commanded the human, saying, "You *may eat* from every tree of the garden. 17. But from the tree of knowledge of good and bad: you shall not eat from it, because in the day you eat from it: you'll *die!*"

18. And YHWH said, "It's not good for the human to be by himself. I'll make for him a strength corresponding to him." 19. And YHWH fashioned from the ground every animal of the field and every bird of the skies and brought it to the human to see what he would call it. And whatever the human would call it, each living being, that would be its name. 20. And the human gave names to every beast and bird of the skies and every animal of the field. But He did not find for the human a strength corresponding to him.

21. And YHWH caused a slumber to descend on the human, and he slept. And He took one of his ribs and closed flesh in its place. 22. And YHWH built the rib that He had taken from the human into a woman and brought her to the human. 23. And the human said, "This time is it: bone of my bones and flesh of my flesh. This will be called 'woman,' for this one was taken from 'man.'"

24. On account of this a man leaves his father and his mother and clings to his woman, and they become one flesh.

25. And the two of them were naked, the human and his woman, and they were not embarrassed.

3:1. And the snake was slier than every animal of the field that YHWH had made, and he said to the woman, "Has God indeed said you may not eat from any tree of the garden?"

2. And the woman said to the snake, "We may eat from the fruit of the trees of the garden. 3. But from the fruit of the tree that is within the garden God has said, 'You shall not eat from it, and you shall not touch it, or else you'll die.'"

4. And the snake said to the woman, "You won't *die!* 5. Because God knows that in the day you eat from it your eyes will be opened, and you'll be like God—knowing good and bad." 6. And the woman saw that the tree was good for eating and that it was an attraction to the eyes, and the tree was desirable to bring about

understanding, and she took some of its fruit, and she ate, and gave to her man with her, and he ate. 7. And the eyes of the two of them were opened, and they knew that they were naked. And they picked fig leaves and made loincloths for themselves.

8. And they heard the sound of YHWH walking in the garden in the wind of the day, and the human and his woman hid from YHWH among the garden's trees. 9. And YHWH called the human and said to him, "Where are you?"

10. And he said, "I heard the sound of you in the garden and was afraid because I was naked, and I hid."

11. And He said, "Who told you that you were naked? Have you eaten from the tree from which I commanded you not to eat?"

12. And the human said, "The *woman,* whom *you* placed with me, *she* gave me from the tree, and I ate."

13. And YHWH said to the woman, "What is this that you've done?"

And the woman said, "The *snake* tricked me, and I ate."

14. And YHWH said to the snake, "Because you did this, you are cursed out of every domestic animal and every animal of the field, you'll go on your belly, and you'll eat dust all the days of your life. 15. And I'll put enmity between you and the woman and between your seed and her seed. He'll strike you at the head, and you'll strike him at the heel."

16. To the woman He said, "I'll make your suffering and your labor pain great. You'll have children in pain. And your desire will be for your man, and he'll dominate you."

17. And to the human He said, "Because you listened to your woman's voice and ate from the tree about which I commanded you saying, 'You shall not eat from it,' the ground is cursed on your account. You'll eat from it with suffering all the days of your life. 18. And it will grow thorn and thistle at you, and you'll eat the field's vegetation. 19. By the sweat of your nostrils you'll eat bread

until you go back to the ground, because you were taken from it; because you are dust and you'll go back to dust."

20. And the human called his woman "Eve," because she was mother of all living.

21. And YHWH made skin garments for the human and his woman and dressed them.

22. And YHWH said, "Here, the human has become like one of us, to know good and bad. And now, in case he'll put out his hand and take from the tree of life as well, and eat and live forever:" 23. And YHWH put him out of the garden of Eden, to work the ground from which he was taken. 24. And He expelled the human, and He had the cherubs and the flame of a revolving sword reside at the east of the garden of Eden to watch over the way to the tree of life.

4:1. And the human knew Eve, his wife, and she became pregnant and gave birth to Cain and said, "I've created a man with YHWH." 2. And she went on to give birth to his brother, Abel. And Abel was a shepherd of flocks, and Cain was a worker of ground. 3. And it was at the end of some days, and Cain brought an offering to YHWH from the fruit of the ground. 4. And Abel brought, as well, from the firstborn of his flock and their fat. And YHWH paid attention to Abel and his offering 5. and did not pay attention to Cain and his offering. And Cain was very angered, and his face was fallen.

6. And YHWH said to Cain, "Why are you angered, and why has your face fallen? 7. Is it not that if you do well you'll be raised, and if you don't do well then sin crouches at the threshold? And its desire will be for you. And you'll dominate it."

8. And Cain said to his brother Abel, "Let's go out to the field." And it was while they were in the field, and Cain rose against Abel, his brother, and killed him.

9. And YHWH said to Cain, "Where is Abel, your brother?" And he said, "I don't know. Am I my brother's watchman?"

10. And He said, "What have you done? The voice of your brother's blood is crying to me from the ground! 11. And now you're cursed from the ground that opened its mouth to take your brother's blood from your hand. 12. When you work the ground it won't continue to give its potency to you. You'll be a roamer and rover in the earth."

13. And Cain said to YHWH, "My crime is greater than I can bear. 14. Here, you've exiled me from the face of the ground today, and I'll be hidden from your presence, and I'll be a roamer and rover in the earth, and anyone who finds me will kill me."

15. And YHWH said to him, "Therefore: anyone who kills Cain, he'll be avenged sevenfold." And YHWH set a sign for Cain so that anyone who finds him would not kill him. 16. And Cain went out from YHWH's presence and lived in the land of roving, east of Eden.

17. And Cain knew his wife, and she became pregnant and gave birth to Enoch. And he was a builder of a city, and he called the name of the city like the name of his son: Enoch. 18. And Irad was born to Enoch, and Irad fathered Mehuya-el, and Mehuya-el fathered Metusha-el, and Metusha-el fathered Lamech. 19. And Lamech took two wives. The one's name was Adah, and the second's name was Zillah. 20. And Adah gave birth to Yabal. He was father of tentdweller and cattleman. 21. And his brother's name was Yubal. He was father of every player of lyre and pipe. 22. And Zillah, too, gave birth to Tubal-Cain, forger of every implement of bronze and iron. And Tubal-Cain's sister was Naamah. 23. And Lamech said to his wives:

> Adah and Zillah, listen to my voice,
> Wives of Lamech, hear what I say.

For I've killed a man for a wound to me
And a boy for a hurt to me,
24. For Cain will be avenged sevenfold
And Lamech seventy-seven.

26b. Then it was begun to invoke the name YHWH.

5:28b. And he fathered a son 29. and called his name Noah, saying, "This one will console us from our labor and from our hands' suffering from the ground, which YHWH has cursed."

6:1. And it was when humankind began to multiply on the face of the ground and daughters were born to them: 2. and the sons of God saw the daughters of humankind, that they were attractive, and they took women, from all they chose. 3. And YHWH said, "My spirit won't stay in humankind forever, since they're also flesh; and their days shall be a hundred twenty years." 4. The Nephilim were in the earth in those days and after that as well, when the sons of God came to the daughters of humankind, and they gave birth by them. They were the heroes of old, people of renown.

5. And YHWH saw that human bad was multiplied in the earth, and every inclination of their heart's thoughts was only bad all the day. 6. And YHWH regretted that He had made humankind in the earth.

And He was grieved to His heart.

7. And YHWH said, "I'll wipe out the human whom I've created from the face of the ground, from human to animal to creeping thing to bird of the skies, because I regret that I made them."

8. And Noah found favor in YHWH's eyes. 7:1. And YHWH said to Noah, "Come, you and all your household, into an ark, for

I've seen you as virtuous before me in this generation. 2. Of all the clean beasts, take seven pairs, man and his woman; and of the beasts that are not clean, two, man and his woman. 3. Also of the birds of the skies seven pairs, male and female, to keep seed alive on the face of the earth. 4. For in seven more days I'll rain on the earth, forty days and forty nights, and I'll wipe out all the substance that I've made from upon the face of the earth."

5. And Noah did according to all that YHWH had commanded him. 7. And Noah and his sons and his wife and his sons' wives with him came to the ark from before the waters of the flood. 10. And seven days later the waters of the flood were on the earth. 12. And there was rain on the earth, forty days and forty nights. 16b. And YHWH closed it for him. 17. And the flood was on the earth for forty days, and the waters multiplied and raised the ark, and it was lifted from the earth. 18. And the waters grew strong and multiplied very much on the earth, and the ark went on the face of the waters. 19. And the waters grew very, very strong on the earth, and they covered all the high mountains that are under all the skies. 20. Fifteen cubits above, the waters grew stronger, and they covered the mountains. 22. Everything that had the breathing spirit of life in its nostrils, everything that was on the ground, died. 23. And He wiped out all the substance that was on the face of the earth, from human to beast to creeping thing and to bird of the skies, and they were wiped out from the earth, and only Noah and those who were with him in the ark were left.

8:2b. And the rain was restrained from the skies. 3a. And the waters went back from on the earth, going back continually. 6. And it was at the end of forty days, and Noah opened the window of the ark that he had made. 8. And he sent out a dove from him to see whether the waters had eased from the face of the earth. 9. And the dove did not find a resting place for its foot, and it came back to him to the ark, for waters were on the face of the earth, and he put out his hand and took it and brought it to him to the ark. 10. And he waited still another seven days, and he again sent out a dove from the ark. 11. And the dove came to him at evening time, and here was an olive leaf torn off in its mouth, and Noah knew that the waters had eased from the earth. 12. And he waited still another seven days, and he sent out a dove, and it did not come back to him ever again. 13b. And Noah turned back the covering of the ark and looked, and here the face of the earth had dried.

20. And Noah built an altar to YHWH, and he took some of each of the clean beasts and of each of the clean birds, and he offered sacrifices on the altar. 21. And YHWH smelled the pleasant smell, and YHWH said to His heart, "I won't curse the ground on account of humankind again, because the inclination of the human heart is bad from their youth, and I won't strike all the living again as I have done. 22. All the rest of the earth's days, seed and harvest, and cold and heat, and summer and winter, and day and night will not cease."

9:18. And Noah's sons who went out from the ark were Shem and Ham and Japheth. And Ham: he was the father of Canaan. 19. These three were Noah's sons, and all the earth expanded from these. 20. And Noah began to be a man of the ground, and he planted a vineyard. 21. And he drank from the wine and was drunk. And he was exposed inside his tent. 22. And Ham, the father of Canaan, saw his father's nakedness and told his two brothers outside. 23. And Shem and Japheth took a garment and put it on both their shoulders and went backwards and covered their father's nakedness. And they faced backwards and did not see their father's nakedness. 24. And Noah woke up from his wine, and he knew what his youngest son had done to him. 25. And he said, "Canaan is cursed. He'll be a servant of servants to his brothers." 26. And he said, "Blessed is YHWH, God of Shem, and may Canaan be a servant to them. 27. May God enlarge Japheth, and may he dwell in the tents of Shem, and may Canaan be a servant to them."

11:1. And all the earth was one language and the same words. 2. And it was when they were traveling from the east: and they found a valley in the land of Shinar and lived there. 3. And they said to one another, "Come on, let's make bricks and fire them." And they had brick for stone, and they had bitumen for mortar. 4. And they said, "Come on, let's build ourselves a city and a tower, and its top will be in the skies, and we'll make ourselves a name, or else we'll scatter over the face of all the earth."

5. And YHWH went down to see the city and the tower that humankind had built. 6. And YHWH said, "Here, they're one people, and they all have one language, and this is what they've begun to do. And now nothing that they'll scheme to do will be precluded from them. 7. Come on, let's go down and babble their language there so that one won't understand another's language." 8. And YHWH scattered them from there over the face of all the earth, and they stopped building the city. 9. On account of this its name was called Babylon, for YHWH babbled the language of all the earth there, and YHWH scattered them from there over the face of all the earth.

12:1. And YHWH said to Abraham, "Go from your land and from your birthplace and from your father's house to the land that I'll show you. 2. And I'll make you into a big nation, and I'll bless you and make your name great. And *be* a blessing! 3. And I'll bless those who bless you, and those who affront you I'll curse. And all the families of the earth will be blessed through you." 4a. And Abraham went as YHWH had spoken to him, and Lot went with him. 6. And Abraham passed through the land as far as the place of Shechem, as far as Alon Moreh. And the Canaanite was in the land then.

7. And YHWH appeared to Abraham and said, "I'll give this land to your seed." And he built an altar there to YHWH who had appeared to him. 8. And he moved on from there to the hill country east of Beth-El and pitched his tent—Beth-El to the west and Ai to the east—and he built an altar to YHWH there and invoked the name YHWH. 9. And Abraham traveled on, going to the Negeb.

10. And there was a famine in the land, and Abraham went down to Egypt to reside there because the famine was heavy in the land. 11. And it was when he was close to coming to Egypt: and he said to Sarah, his wife, "Here, I know that you're a beautiful woman. 12. And it will be when the Egyptians see you that they'll say, 'This is his wife,' and they'll kill me and keep you alive. 13. Say you're my sister so it will be good for me on your account and I'll stay alive because of you." 14. And it was as Abraham came to Egypt, and the Egyptians saw the woman, that she was very beautiful. 15. And Pharaoh's officers saw her and praised her to Pharaoh, and the woman was taken to Pharaoh's house. 16. And he was good to Abraham on her account, and he had a flock and oxen and he-asses and servants and maids and she-asses and camels. 17. And YHWH plagued Pharaoh and his house, big plagues, over the matter of Sarah, Abraham's wife.

18. And Pharaoh called Abraham and said, "What is this you've done to me? Why didn't you tell me that she was your wife? 19. Why did you say, 'She's my sister,' so I took her for a wife for myself! And now, here's your wife. Take her and go." 20. And Pharaoh commanded people over him and sent him and his wife and all he had away.

13:1. And Abraham went up from Egypt, he and his wife and all he had and Lot with him, to the Negeb. 2. And Abraham was very heavy with cattle and silver and gold. 3. And he went on his journeys from the Negeb as far as Beth-El, as far as the place where his tent was at the start, between Beth-El and Ai, 4. to the place of the altar that he had made at the beginning. And Abraham invoked the name YHWH there. 5. And Lot, who was going with Abraham, also had a flock and oxen and tents. 7. And there was a quarrel between those who herded Abraham's cattle and those who herded Lot's cattle. And the Canaanite and the Perizzite lived in the land then. 8. And Abraham said to Lot, "Let there be no strife between me and you and between my herders and your herders, because we're brothers. 9. Isn't the whole land before you? Separate from me: if left then I'll go right, and if right then I'll go left." 10. And Lot raised his eyes and saw all the plain of the Jordan, that it was well watered (before YHWH's destroying Sodom and Gomorrah) like YHWH's garden, like the land of Egypt, as you come to Zoar. 11a. And Lot chose all the plain of the Jordan for himself, and Lot traveled east 12b. and tented as far as Sodom. 13. And the people of Sodom were very bad and sinful to YHWH.

14. And YHWH said to Abraham after Lot's separation from him, "Lift your eyes and see from the place where you are to the north and south and east and west. 15. For all the land that you see, I'll give it to you and to your seed forever. 16. And I'll make your seed like the dust of the land, so that if a man could count the dust of the land then your seed also could be counted. 17. Get up, go around in the land to its length and its breadth, because I'm giving it to you." 18. And Abraham took up his tent and came and lived among the oaks of Mamre which are in Hebron, and he built an altar to YHWH there.

15:6. And he had faith in YHWH, and He reckoned it to him as virtue. 7. And He said to him, "I am YHWH, who brought you out of Haran to give you this land, to possess it."

8. And he said, "My Lord YHWH, how will I know that I'll possess it?"

9. And He said to him, "Take a three-year-old heifer and a three-year-old she-goat and a three-year-old ram and a dove and a pigeon for me."

10. And he took all of these for Him and split them up the middle and set each half opposite its other half, but he did not split the birds. 11. And birds of prey came down on the carcasses, and Abraham drove them away. 12. And the sun was about to set, and a slumber had descended on Abraham; and, here, a big, dark terror was descending on him. 17. And the sun was setting, and there was darkness, and here was an oven of smoke and a flame of fire that went between these pieces. 18. In that day YHWH made a covenant with Abraham, saying, "I've given this land to your seed, from the river of Egypt to the big river, the river Euphrates: 19. the Cainites and the Kenizzites and the Kadmonites 20. and the Hittites and the Perizzites and the Rephaim 21. and the Amorites and the Canaanites and the Girgashites and the Jebusites."

16:1. And Abraham's wife, Sarah, had not given birth by him. And she had an Egyptian maid, and her name was Hagar. 2. And Sarah said to Abraham, "Here, YHWH has held me back from giving birth. Come to my maid. Maybe I'll get 'childed' through her." And Abraham listened to Sarah's voice. 4. And he came to Hagar, and she became pregnant. And she saw that she had become pregnant, and her mistress was lowered in her eyes.

5. And Sarah said to Abraham, "My injury is on you. I, *I,* placed my maid in your bosom, and she saw that she had become pregnant, and I was lowered in her eyes. Let YHWH judge between me and you."

6. And Abraham said to Sarah, "Here, your maid is in your hand. Do whatever is good in your eyes to her." And Sarah degraded her, and she fled from her.

7. And an angel of YHWH found her at a pool of water in the wilderness, by the pool on the way to Shur, 8. and said, "Hagar, Sarah's maid, from where have you come, and where will you go?"

And she said, "I'm fleeing from Sarah, my mistress."

9. And the angel of YHWH said to her, "Go back to your mistress, and suffer the degradation under her hands." 10. And the angel of YHWH said, "I'll *multiply* your seed, and it won't be countable because of its great number." 11. And the angel of YHWH said to her, "Here, you're pregnant and will give birth to a son, and you shall call his name Ishmael, for YHWH has listened to your suffering. 12. And he'll be a wild ass of a man, his hand against everyone, and everyone's hand against him, and he'll tent among all his brothers."

13. And she called the name of YHWH who spoke to her "You are El-roi," for she said, "Have I also seen after the one who sees me here?" 14. On account of this the well was called Beer-lahai-roi. Here it is between Kadesh and Bered.

18:1. And YHWH appeared to him at the oaks of Mamre. And he was sitting at the tent entrance in the heat of the day, 2. and he raised his eyes and saw, and here were three people standing over him. And he saw and ran toward them from the tent entrance and bowed to the ground. 3. And he said, "My Lord, if I've found favor in your eyes don't pass on from your servant. 4. Let a little water be gotten, and wash your feet and relax under a tree, 5. and let me get a bit of bread, and satisfy your heart. Afterward you'll pass on, for that's why you've passed by your servant."

And they said, "Do that, as you've spoken."

6. And Abraham hurried to the tent, to Sarah, and said, "Hurry, three measures of fine flour. Knead it and make cakes." 7. And Abraham ran to the herd and took a calf, tender and good, and he gave it to a servant, and he hurried to prepare it. 8. And he took curds and milk and the calf that he had prepared and placed them in front of them, and he was standing over them under the tree, and they ate.

9. And they said to him, "Where is Sarah, your wife?"

And he said, "Here in the tent."

10. And He said, "I shall *come back* to you at the time of life, and, here, Sarah, your wife, will have a son."

And Sarah was listening at the tent entrance, which was behind him. 11. And Abraham and Sarah were old, well along in days; Sarah had stopped having the way of women. 12. And Sarah laughed inside her and said, "After I've withered am I to have pleasure? And my lord is old!"

13. And YHWH said to Abraham, "Why is this? Sarah laughed, saying, 'Shall I indeed give birth? And I am old!' Is anything too wondrous for YHWH? 14. At the appointed time I'll come back to you, at the time of life, and Sarah will have a son."

15. And Sarah lied, saying, "I didn't laugh," because she was afraid.

And He said, "No, but you did laugh."

16. And the people got up from there and gazed at the sight of Sodom, and Abraham was going with them to send them off. 17. And YHWH said, "Shall I conceal what I'm doing from Abraham,

18. since Abraham will become a big and powerful nation, and all the nations of the earth will be blessed through him? 19. For I've known him for the purpose that he'll command his children and his house after him, and they'll observe YHWH's way, to do virtue and judgment, and for the purpose of YHWH's bringing upon Abraham what He spoke about him."

20. And YHWH said, "The cry of Sodom and Gomorrah—how great it is; and their sin—how very heavy it is. 21. Let me go down, and I'll see if they've done, all told, like the cry that has come to me. And if not, let me know." 22. And the people turned from there and went to Sodom.

And Abraham was still standing before YHWH, 23. and Abraham came over and said, "Will you also annihilate the virtuous with the wicked? 24. Maybe there are fifty virtuous people in the city. Will you also annihilate and not sustain the place for the fifty virtuous who are in it? 25. Far be it from you to do a thing like this, to kill virtuous with wicked—and it will be the same for the virtuous and the wicked—far be it from you. Will the judge of all the earth not do justice?"

26. And YHWH said, "If I find in Sodom fifty virtuous people in the city then I'll sustain the whole place for their sake."

27. And Abraham answered, and he said, "Here I've dared to speak to my Lord, and I'm dust and ashes. 28. Maybe the fifty virtuous people will be short by five. Will you destroy the whole city for the five?"

And He said, "I won't destroy if I find forty-five there."

29. And he went on again to speak to Him and said, "Maybe forty will be found there."

And He said, "I won't do it for the sake of the forty."

30. And he said, "May my Lord not be angry, and let me speak. Maybe thirty will be found there."

And He said, "I won't do it if I find thirty there."

31. And he said, "Here I've dared to speak to my Lord. Maybe twenty will be found there."

And He said, "I won't destroy for the sake of the twenty."

32. And he said, "May my Lord not be angry, and let me speak just this one time. Maybe ten will be found there."

And He said there, "I won't destroy for the sake of the ten."

33. And YHWH went when He had finished speaking to Abraham, and Abraham went back to his place.

19:1. And the two angels came to Sodom in the evening, and Lot was sitting at Sodom's gate, and Lot saw and got up toward them, and he bowed, nose to the ground. 2. And he said, "Here, my lords, turn to your servant's house and spend the night and wash your feet, and you'll get up early and go your way."

And they said, "No, we'll spend the night in the square." 3. And he pressed them very much, and they turned to him and came to his house, and he made a feast and baked unleavened bread for them, and they ate.

4. They had not yet lain down, and the people of the city, the people of Sodom, surrounded the house, from youth to old man, all the people, from the farthest reaches. 5. And they called to Lot and said to him, "Where are the people who came to you tonight? Bring them out to us, and let's know them!"

6. And Lot went out to them at the entrance and closed the door behind him, 7. and he said, "Don't do bad, my brothers." 8. Here I have two daughters who haven't known a man. Let me bring them out to you, and do to them as is good in your eyes. Only don't do anything to these people, because that is why they came under the shadow of my roof."

9. And they said, "Come over here," and they said, "This one comes to reside, and then he *judges!* Now we'll be worse to you

than to them." And they pressed the man, Lot, very much and came over to break the door. 10. And the people reached their hand out and brought Lot in to them in the house, and they closed the door. 11. And they struck the people who were at the house's entrance with blindness, from smallest to biggest, and they wearied themselves with finding the entrance.

12. And the people said to Lot, "Who else do you have here—son-in-law and your sons and your daughters and all that you have in the city—take them out from the place, 13. because we're destroying this place, because its cry has grown big before YHWH's face, and YHWH has sent us to destroy it."

14. And Lot went out and spoke to his sons-in-law, who had married his daughters, and said, "Get up, get out of this place, because YHWH is destroying the city." And he was like a joker in his sons-in-law's eyes.

15. And as the dawn rose the angels urged Lot, saying, "Get up, take your wife and your two daughters who are present, or else you'll be annihilated for the city's crime." 16. And he delayed. And the people took hold of his hand and his wife's hand and his two daughters' hands because of YHWH's compassion for him, and they brought him out and set him outside of the city.

17. And it was as they were bringing them out, and He said, "Escape for your life. Don't look behind you and don't stop in all of the plain. Escape to the mountain or else you'll be annihilated."

18. And Lot said to them, "Let it not be, my Lord. 19. Here, your servant has found favor in your eyes, and you've magnified your kindness that you've done for me, keeping my soul alive, and I'm not able to escape to the mountain, in case the bad thing will cling to me and I'll die. 20. Here, this city is close to flee there, and it's small. Let me escape there—isn't it small?—and my soul will live."

21. And He said to him, "Here, I've granted you this thing, too, that I won't overturn the city of which you spoke. 22. Quickly, escape there, because I can't do a thing until you get there."

On account of this the city's name was called Zoar.

23. The sun rose on the earth, and Lot was coming to Zoar. 24. And YHWH rained brimstone and fire on Sodom and on Gomorrah, from YHWH out of the skies. 25. And He overturned these cities and all of the plain and all of the inhabitants of the cities and all the growth of the ground. 26. And his wife looked behind him, and she was a pillar of salt!

27. And Abraham got up early in the morning to the place where he had stood in YHWH's presence, 28. and he gazed at the sight of Sodom and Gomorrah and at the sight of all the land of the plain. And he saw: and, here, the smoke of the land went up like the smoke of a furnace.

30. And Lot went up from Zoar and lived in the mountain, and his two daughters with him, because he feared to live in Zoar, and he lived in a cave, he and his two daughters. 31. And the firstborn said to the younger one, "Our father is old, and there's no man in the earth to come to us in the way of all the earth. 32. Come on, let's make our father drink wine, and let's lie with him and make seed live from our father." 33. And they made their father drink wine in that night, and the firstborn came and lay with her father. And he did not know of her lying down and her getting up. 34. And it was on the next day, and the firstborn said to the younger one, "Here, I lay with my father last night. Let's make him drink wine tonight as well, and you come and lie with him, and we'll make seed live from our father." 35. And they made their father drink wine in that night as well, and the younger one got up and lay with him, and he did not know of her lying down and her getting up. 36. And Lot's two daughters became pregnant by their

father. 37. And the firstborn gave birth to a son and called his name Moab. He is the father of Moab to this day. 38. And the younger one, she too gave birth to a son and called his name ben-Ammi. He is the father of the children of Ammon to this day.

21:1a. And YHWH took account of Sarah as He had said, 2a. and Sarah became pregnant and gave birth to a son for Abraham in his old age. 7. And she said, "Who would have said to Abraham, 'Sarah has nursed children'? Yet I've given birth to a son in his old age."

22:20. And it was after these things, and it was told to Abraham, saying, "Here Milcah has given birth, she also, to sons for your brother Nahor: 21. Uz, his firstborn, and Buz his brother, and Kemuel, the father of Aram, 22. and Chesed and Hazo and Pildash and Jidlaph and Bethuel. 23. And Bethuel fathered Rebekah." Milcah gave birth to these eight for Nahor, Abraham's brother. 24. And his concubine, whose name was Reumah, she too gave birth, to Tebah and Gaham and Tahash and Maacah. 24:1. And Abraham was old, well along in days, and YHWH had blessed Abraham in everything. 2. And Abraham said to his servant, the elder of his house, who dominated all that he had, "Place your hand under my thigh, 3. and I'll have you swear by YHWH, God of the skies and God of the earth, that you won't take a wife for my son from the daughters of the Canaanite among whom I live, 4. but you'll go to my land and my birthplace and take a wife for my son, for Isaac."

5. And the servant said to him, "Maybe the woman won't be willing to follow me to this land. Shall I *take* your son *back* to the land you came from?"

6. And Abraham said to him, "Watch yourself, that you don't take my son back there. 7. YHWH, God of the skies, who took me

from my father's house and from the land of my birth and who spoke to me and who swore to me, saying, 'I'll give this land to your seed,' He'll send His angel ahead of you, and you shall take a wife for my son from there. 8. And if the woman won't be willing to go after you then you'll be freed from this oath of mine. Only you are not to take my son back there."

9. And the servant placed his hand under his lord Abraham's thigh and swore to him about this thing. 10. And the servant took ten camels of his lord's camels and went, and all of his lord's best things were in his hand. And he got up and went to Aram Naharaim, to the city of Nahor, 11. and he had the camels kneel outside the city at the water well at evening time, the time that the women went out to draw water. 12. And he said, "YHWH, God of my lord Abraham, make something happen before me today and show kindness to my lord Abraham: 13. Here I am, standing over the pool of water, and the daughters of the people of the city are going out to draw water. 14. And it will be that the girl to whom I'll say, 'Tip your jar and let me drink,' and she'll say, 'Drink, and I'll water your camels, too,' she'll be the one you've designated for your servant, for Isaac, and I'll know by this that you've shown kindness to my lord."

15. And he had not even finished speaking, and here was Rebekah—who was born to Bethuel son of Milcah, wife of Nahor, Abraham's brother—coming out, and her jar was on her shoulder. 16. And the girl was very good looking, a virgin, and no man had known her. And she went down to the pool and filled her jar and went up. 17. And the servant ran toward her and said, "Give me a little water from your jar."

18. And she said, "Drink, my lord," and she hurried and lowered her jar on her arm and let him drink.

19. And she finished letting him drink, and she said, "I'll draw water for your camels, too, until they finish drinking." 20. And she hurried and emptied her jar into the trough and ran again to the well to draw water, and she drew water for all his camels. 21. And the man, astonished at her, was keeping quiet so as to know if YHWH had made his trip successful or not.

22. And it was when the camels finished drinking: and the man took a gold ring—its weight was a half shekel—and two bracelets on her hands—their weight was ten of gold. 23. And he said, "Tell me, whose daughter are you? Is there a place at your father's house for us to spend the night?"

24. And she said to him, "I'm a daughter of Bethuel, Milcah's son whom she bore to Nahor." 25. And she said to him, "We also have plenty of both straw and fodder, also a place to spend the night."

26. And the man knelt and bowed to YHWH 27. and said, "Blessed is YHWH, God of my lord Abraham, whose kindness and faithfulness have not left my lord. YHWH has led me to the house of my lord's brother."

28. And the girl ran and told her mother's household about these things. 29. And Rebekah had a brother, and his name was Laban, and Laban ran outside to the man, at the pool. 30. And it was when he saw the ring and the bracelets on his sister's hands and when he heard his sister Rebekah's words, saying, "The man spoke like this to me." And he came to the man, and here he was standing by the camels at the pool. 31. And he said, "Come, blessed one of YHWH. Why do you stand outside when I've prepared the house and a place for the camels!" 32. And the man came to the house and unloaded the camels and gave straw and fodder to the camels and water to wash his feet and the feet of the people who were with him. 33. And [bread] was set in front of him to eat.

And he said, "I won't eat until I've said what I have to say."

And he said, "Speak."

34. And he said, "I am Abraham's servant. 35. And YHWH has blessed my lord very much, and he has become great. And He has given him a flock and oxen and silver and gold and male and female servants and camels and asses. 36. And my lord's wife, Sarah, gave birth to a son by my lord in her old age, and he has given him everything that he has. 37. And my lord had me swear, saying, 'You shall not take a wife for my son from the daughters of the Canaanite in whose land I live, 38. but you shall go to my father's house and to my family and take a wife for my son.' 39. And I said to my lord, 'Maybe the woman won't follow me.' 40. And he said to me, 'YHWH, before whom I have walked, will send His angel with you and make your trip successful, and you shall take a wife for my son from my family and from my father's house. 41. Then you'll be freed from my oath: when you come to my family, and if they won't give her to you, and you'll be free from my oath.' 42. And I came to the pool today, and I said, 'YHWH, God of my lord Abraham, if you're making my trip on which I'm going successful, 43. here I am, standing at the pool of water, and it will be that the young woman who goes to draw water and I say to her, "Let me drink a little water from your jar," 44. and she says to me, "Both drink yourself and I'll draw water for your camels, too," she will be the woman whom YHWH has designated for my lord's son.' 45. I hadn't even finished speaking in my heart, and here was Rebekah coming out, and her jar was on her shoulder, and she went down to the pool and drew water. And I said to her, 'Give me a drink.' 46. And she hurried and lowered her jar from on her and said, 'Drink, and I'll water your camels, too.' And I drank, and she watered the camels, too. 47. And I asked her and said, 'Whose daughter are

you?' And she said, 'The daughter of Bethuel son of Nahor, whom Milcah bore for him.' And I put the ring on her nose and the bracelets on her hands, 48. and I knelt and bowed to YHWH and blessed YHWH, my lord Abraham's God, who had led me in a faithful way to take a daughter of my lord's brother for his son. 49. And now, if you're exercising kindness and faithfulness with my lord, tell me; and if not, tell me; so I'll turn to right or to left."

50. And Laban and Bethuel answered, and they said, "The thing has come from YHWH. We can't speak bad or good to you. 51. Here is Rebekah before you. Take her and go, and let her be a wife to your lord's son as YHWH has spoken." 52. And it was, when Abraham's servant heard their words, that he bowed to the ground to YHWH. 53. And the servant brought out silver articles and gold articles and garments and gave them to Rebekah and gave precious things to her brother and to her mother. 54. And they ate and drank, he and the people who were with him, and spent the night.

And they got up in the morning, and he said, "Send me back to my lord."

55. And her brother and her mother said, "Let the girl stay with us a few days—or ten. After that, she'll go."

56. And he said to them, "Don't hold me back, since YHWH has made my trip successful. Send me away, so I may go to my lord."

57. And they said, "We'll call the girl and ask her from her own mouth." 58. And they called Rebekah and said to her, "Will you go with this man?"

And she said, "I'll go."

59. And they sent their sister Rebekah and her nurse and Abraham's servant and his people. 60. And they blessed Rebekah and said to her, "You're our sister. Become thousands of ten thousands. And may your seed possess his enemies' gate." 61. And

Rebekah and her maids got up and rode on the camels and followed the man, and the servant took Rebekah and went.

62. And Isaac was coming from the area of Beer-lahai-roi, and he was living in the territory of the Negeb, 63. and Isaac went to meditate in a field toward evening, and he raised his eyes and saw, and here were camels coming. 64. And Rebekah raised her eyes and saw Isaac, and she fell from the camel. 65. And she said to the servant, "Who is that man who's walking in the field toward us?"

And the servant said, "He's my lord." And she took a veil and covered herself.

66. And the servant told Isaac all the things that he had done. 67. And Isaac brought her to his mother Sarah's tent. And he took Rebekah, and she became his wife, and he loved her.

And Isaac was consoled after his mother.

25:8a. And Abraham died at a good old age. 11b. And Isaac lived at Beer-lahai-roi. 21. And Isaac prayed to YHWH for his wife because she was infertile, and YHWH was prevailed upon by him, and his wife Rebekah became pregnant. 22. And the children struggled inside her, and she said, "If it's like this, why do I exist?" And she went to inquire of YHWH.

23. And YHWH said to her,

> Two nations are in your womb,
> and two peoples will be separated from your insides,
> and one people will be mightier than the other people,
> and the older the younger will serve.

24. And her days to give birth were completed, and here were twins in her womb. 25. And the first came out all ruddy, like a

hairy robe, and they called his name Esau. 26. And after that his brother came out, and his hand was holding Esau's heel, and he called his name Jacob. And Isaac was sixty years old at their birth.

27. And the boys grew up, and Esau was a man who knew hunting, a man of the field, and Jacob was a simple man, living in tents. 28. And Isaac loved Esau because he put game meat in his mouth, and Rebekah loved Jacob.

29. And Jacob made a stew, and Esau came from the field, and he was exhausted. 30. And Esau said to Jacob, "Will you feed me some of the red stuff, this red stuff, because I'm exhausted." (On account of this he called his name "Edom.")

31. And Jacob said, "Sell your birthright to me, today."

32. And Esau said, "Here I'm going to die, and what use is this, that I have a birthright?"

33. And Jacob said, "Swear to me, today."

And he swore to him and sold his birthright to Jacob. 34. And Jacob gave bread and lentil stew to Esau. And he ate and drank and got up and went. And Esau disdained the birthright.

26:1. And there was a famine in the land (other than the first famine, which was in Abraham's days), and Isaac went to Abimelek, king of the Philistines, at Gerar. 2. And YHWH appeared to him and said, "Don't go down to Egypt. Tent in the land that I say to you. 3. Reside in this land, and I'll be with you and bless you, for I'll give all these lands to you and your seed, and I'll uphold the oath that I swore to Abraham your father, 4. and I'll multiply your seed like the stars of the skies and give to your seed all these lands, and all the nations of the earth will be blessed through your seed 5. because Abraham listened to my voice and kept my watch, my commandments, my laws, and my instructions." 6. And so Isaac lived in Gerar.

7. And the people of the place asked about his wife, and he said, "She's my sister," because he was afraid to say, "My wife," or else "the people of the place will kill me for Rebekah, because she's good looking." 8. And it was when his days grew long there: and Abimelek, king of the Philistines, gazed through the window, and he saw: and here was Isaac "fooling around" with Rebekah his wife!

9. And Abimelek called Isaac and said, "Really, here, she's your wife, and how could you say, 'She's my sister'?"

And Isaac said to him, "Because I said, 'Or else I'll die over her.'"

10. And Abimelek said, "What is this you've done to us? One of the people nearly could have lain with your wife, and you would have brought guilt on us." 11. And Abimelek commanded all the people saying, "He who touches this man or his wife will be put to *death!*"

12. And Isaac planted seed in that land, and he harvested a hundredfold in that year. And YHWH blessed him, 13. and the man became great, and he went on getting greater until he was very great 14. and had possession of flocks and possession of herds and a large number of servants. And the Philistines envied him. 15. And the Philistines stopped up all the wells that his father's servants had dug in his father Abraham's days, and they covered them with dirt. 16. And Abimelek said to Isaac, "Go from among us, because you've become much more powerful than we are." 17. And Isaac went from there and camped in the wadi of Gerar and lived there. 18. And Isaac went back and dug the water wells which they had dug in the days of his father Abraham and which the Philistines had stopped up after Abraham's death, and he called them by names, like the names that his father had called them. 19. And Isaac's servants dug in the wadi and found a well of fresh water there. 20. And the shepherds of Gerar quarreled with Isaac's shepherds, saying, "The water is ours."

And he called the name of the well Esek because they tangled with him. 21. And they dug another well, and they quarreled over it also, and he called its name Sitnah. 22. And he moved on from there and dug another well, and they did not quarrel over it, and he called its name Rehovot and said, "Because now YHWH has extended for us, and we've been fruitful in the land."

23. And he went up from there to Beer-sheba. 24. And YHWH appeared to him that night and said, "I'm your father Abraham's God. Don't be afraid, because I'm with you, and I'll bless you and multiply your seed on account of Abraham, my servant." 25. And he built an altar there and invoked the name YHWH. And he pitched his tent there, and Isaac's servants dug a well there.

26. And Abimelek and Ahuzat, his companion, and Phicol, the commander of his army, went to him from Gerar. 27. And Isaac said to them, "Why have you come to me, since you hated me and sent me away from you?"

28. And they said, "We've seen that YHWH was with you; and we say: Let there be a pledge between us, between us and you, and let us make a covenant with you 29. that you won't do bad toward us as we haven't touched you and as we've done only good toward you and sent you away in peace. You are now blessed by YHWH." 30. And he made a feast for them, and they ate and drank. 31. And they got up early in the morning and swore, each man to his brother, and Isaac sent them away, and they went from him in peace.

32. And it was in that day, and Isaac's servants came and told him about the well that they had dug and said to him, "We found water." 33. And he called it Seven. On account of this the name of the city is Beer-sheba to this day.

27:1. And it was when Isaac was old and his eyes were too dim for seeing: and he called Esau, his older son, and said to him, "My son."

And he said to him, "I'm here."

2. And he said, "Here, I've become old. I don't know the day of my death. 3. And now, take up your implements: your quiver and your bow, and go out to the field and hunt me some game 4. and make me the kind of delicacies that I love and bring them to me and let me eat, so my soul will bless you before I die."

5. And Rebekah was listening as Isaac was speaking to Esau, his son. And Esau went to the field to hunt game to bring to his father. 6. And Rebekah said to Jacob, her son, "Here I've heard your father speaking to Esau, your brother, saying, 7. 'Bring me game and make me delicacies and let me eat, and I'll bless you in the presence of YHWH before my death.' 8. And now, my son, listen to my voice, to what I'm commanding you. 9. Go to the flock and take two good goat kids for me from there, and I'll make them the kind of delicacies that your father loves. 10. And bring them to your father, and he'll eat, so he'll bless you before his death."

11. And Jacob said to Rebekah, his mother, "Here, Esau, my brother, is a hairy man and I'm a smooth man. 12. Maybe my father will feel me, and I'll be like a trickster in his eyes, and I'll bring a curse on me and not a blessing."

13. And his mother said to him, "Let your curse be on me, my son; just listen to my voice and go take them for me." 14. And he went and took and brought them to his mother, and his mother made the kind of delicacies that his father loved. 15. And Rebekah took the finest clothes of Esau, her older son, that were with her in the house and put them on Jacob, her younger son, 16. and put the skins of the goat kids on his hands and on the smooth part of his neck. 17. And she put the delicacies and the bread that she had made in the hand of Jacob, her son. 18. And he came to his father.

And he said, "My father."

And he said, "I'm here. Who are you, my son?"

19. And Jacob said to his father, "I'm Esau, your firstborn. I did as you spoke to me. Get up, sit, and eat some of my game so your soul will bless me."

20. And Isaac said to his son, "What's this? You were quick to find it, my son."

And he said, "Because YHWH your God made it occur for me."

21. And Isaac said to Jacob, "Come over, and I'll feel you, my son: are you the one, my son Esau, or not?" 22. And Jacob came over to Isaac, his father, and he felt him. And he said, "The voice is the voice of Jacob, and the hands are the hands of Esau." 23. And he did not recognize him because his hands were hairy like the hands of Esau, his brother, and he blessed him. 24. And he said, "You're the one, my son Esau?"

And he said, "I am."

25. And he said, "Bring it over to me and let me eat some of my son's game so that my soul will bless you." And he brought it over to him, and he ate; and he brought him wine, and he drank. 26. And Isaac, his father, said to him, "Come over and kiss me, my son." 27. And he came over and kissed him, and he smelled the aroma of his clothes and blessed him and said, "See, my son's aroma is like the aroma of a field that YHWH has blessed. 28. And may God give you from the dew of the skies and from the fat of the earth and much grain and wine. 29. May peoples serve you and nations bow to you. Be your brothers' superior, and may your mother's sons bow to you. May those who curse you be cursed and those who bless you be blessed."

30. And it was as Isaac finished blessing Jacob, and it was just as Jacob had gone out from the presence of Isaac, his father!—and Esau, his brother, came from his hunting. 31. And he, too, had

made delicacies and brought them to his father. And he said to his father, "Let my father get up and eat some of his son's game so you yourself will bless me."

32. And Isaac, his father, said to him, "Who are you?"

And he said, "I'm your son, your firstborn, Esau."

33. And Isaac had a very big trembling. And he said, "Who? Where is the one who hunted game and brought it to me, and I ate some of it all before you came, and I blessed him? He will in fact be blessed."

34. When Esau heard his father's words he gave a very big and bitter cry. And he said to his father, "Bless me, also me, my father."

35. And he said, "Your brother came with deception, and he took your blessing."

36. And he said, "Was his name really called Jacob! And he's usurped me two times now. He's taken my birthright, and, here, now he's taken my blessing." And he said, "Haven't you saved a blessing for me?"

37. And Isaac answered, and he said to Esau, "Here I've made him your superior, and I've given all his brothers as servants, and I've endowed him with grain and wine. And for you: where, what, will I do, my son?"

38. And Esau said to his father, "Is it one blessing that you have, my father? Bless me, also me, my father." And Esau raised his voice and wept.

39. And Isaac, his father, answered, and he said to him, "Here, away from the fat of the earth will be your home, and from the dew of the skies from above. 40. And you'll live by your sword. And you'll serve your brother. And it will be that when you get dominion you'll break his yoke from your neck."

41. And Esau despised Jacob because of the blessing that his father gave him, and Esau said in his heart, "The days of mourning for my father will be soon, and then I'll kill Jacob, my brother."

42. And the words of Esau, her older son, were told to Rebekah, and she sent and called to Jacob, her younger son, and said to him, "Here, Esau, your brother, consoles himself regarding you with the idea of killing you. 43. And now, my son, listen to my voice and get up, flee to Laban, my brother, at Haran 44. and live with him for a number of days until your brother's fury will turn back, 45. until your brother's anger turns back from you and he forgets what you did to him. And I'll send and take you from there. Why should I be bereaved of the two of you as well in one day?" 28:10. And Jacob left Beer-sheba and went to Haran.

11a. And he happened upon a place and stayed the night there because the sun was setting. 13. And here was YHWH standing over him, and He said, "I am YHWH, your father Abraham's God and Isaac's God. The land on which you're lying: I'll give it to you and to your seed. 14. And your seed will be like the dust of the earth, and you'll expand to the west and east and north and south, and all the families on the land will be blessed through you and through your seed. 15. And here I am with you, and I'll watch over you everywhere that you'll go, and I'll bring you back to this land, for I won't leave you until I've done what I've spoken to you."

16. And Jacob woke from his sleep and said, "YHWH is actually in this place, and I didn't know!" 19. And he called that place's name Beth-El, though in fact Luz was the name of the city at first.

29:1. And Jacob lifted his feet and went to the land of the people of the east. 2. And he looked, and here was a well in the field, and here were three flocks of sheep lying by it, because they watered the flocks from that well, and the stone on the mouth of the well was big, 3. and all the flocks would be gathered there, and they would roll the stone from the mouth of the well and would water the sheep and then would put the stone back in its place on the mouth of the well. 4. And Jacob said to them, "My brothers, where are you from?"

And they said, "We're from Haran."

5. And he said to them, "Do you know Laban, son of Nahor?"

And they said, "We know."

6. And he said to them, "Is he well?

And they said, "Well. And here's Rachel, his daughter, coming with the sheep."

7. And he said, "Here, it will still be daytime for a long time, not the time for gathering the cattle. Water the sheep and go, pasture them."

8. And they said, "We can't until all the flocks will be gathered and they'll roll the stone from the mouth of the well, and then we'll water the sheep."

9. He was still speaking with them, and Rachel came with her father's sheep, because she was a shepherdess. 10. And it was when Jacob saw Rachel, daughter of Laban, his mother's brother, and the sheep of Laban, his mother's brother: and he went over and rolled the stone from the mouth of the well and watered the sheep of Laban, his mother's brother. 11. And Jacob kissed Rachel and raised his voice and wept. 12. And Jacob told Rachel that he was her father's kin and that he was Rebekah's son, and she ran and told her father. 13. And it was, when Laban heard the news of Jacob, his sister's son, that he ran to him and embraced him and kissed him and brought him to his house. And he told Laban all these things. 14. And Laban said to him, "You are indeed my bone and my flesh."

And he stayed with him a month. 15. And Laban said to Jacob, "Is it right because you're my brother that you should work for me for free? Tell me what your pay should be." 16. And Laban had two daughters. The older one's name was Leah, and the younger one's name was Rachel. 17. And Leah's eyes were tender, and Rachel had an attractive figure and was beautiful. 18. And Jacob loved Rachel.

And he said, "I'll work for you seven years for Rachel, your younger daughter."

19. And Laban said, "Better for me to give her to you than for me to give her to another man. Live with me."

20. And Jacob worked seven years for Rachel, and they were like a few days in his eyes because of his loving her. 21. And Jacob said to Laban, "Give me my wife, because my days have been completed, and let me come to her." 22. And Laban gathered all the people of the place and made a feast.

23. And it was in the evening, and he took Leah, his daughter, and brought her to him. And he came to her.

24. And Laban gave her his maid Zilpah—to Leah, his daughter, as a maid.

25. And it was in the morning, and here she was: Leah! And he said to Laban, "What is this you've done to me? Didn't I work with you for Rachel? And why have you deceived me?"

26. And Laban said, "It's not done like that in our place, to give the younger one before the firstborn. 27. Complete this week, and this one will be given to you as well—for the work that you'll do with me: another seven years."

28. And Jacob did so; and he completed this week, and he gave him Rachel, his daughter, for a wife. 29. And Laban gave his maid Bilhah to Rachel, his daughter—to her as a maid. 30. And he also came to Rachel. And he also loved Rachel more than Leah. And he worked with him another seven years.

31. And YHWH saw that Leah was hated, and He opened her womb, and Rachel was infertile. 32. And Leah became pregnant and gave birth to a son and called his name Reuben because she said, "Because YHWH looked at my suffering, so that now my man will love me." 33. And she became pregnant again and gave birth to a son and said, "Because YHWH listened because I was hated and gave me this one, too." And she called his name Simeon. 34. And she became pregnant again and gave birth to a son, and she said, "Now, this time my man will become bound to me because I've given birth to three sons for him." On account of this his name was called Levi. 35. And she became pregnant again and gave birth to a son and said, "This time I'll praise YHWH." On account of this she called his name Judah. And she stopped giving birth.

[Accounts of the births of the other sons, except Joseph, are not present.]

30:24b. [Rachel gave birth to a son and named him Joseph] saying, "May YHWH add another son to me." 25. And it was when Rachel had given birth to Joseph: and Jacob said to Laban, "Send me away and let me go to my place and my land. 26. Give me my wives and my children for whom I've worked for you and let me go; because *you* know my work that I've done for you."

27. And Laban said to him, "If I've found favor in your eyes, I've divined that YHWH has blessed me for your sake," 28. and he said, "Set your wage for me, and I'll pay it."

29. And he said to him, "*You* know how I've worked for you and how your cattle has become with me, 30. that a little that you had

before me expanded into a lot, and YHWH blessed you wherever I set foot. And now, when shall *I* also take care of *my* house?"

31. And he said, "What shall I give you?"

And Jacob said, "You won't *give* me anything—if you'll do this thing for me: Let me go back; I'll tend and watch over your flock. 32. I'll pass among all your flock today removing from there every speckled and spotted lamb and every brown lamb among the sheep and every spotted and speckled one among the goats—and that will be my wage. 33. And my integrity will answer for me in a future day: when you come upon my wage, it will be before you. Any one that isn't speckled and spotted among the goats and brown among the sheep, it's stolen with me."

34. And Laban said, "Here, let it be according to your word." 35. And in that day he removed the he-goats that were streaked and spotted and the she-goats that were speckled and spotted, every one that had white in it, and every brown one among the sheep, and set them in his sons' hand 36. and put three days' distance between him and Jacob. And Jacob was tending Laban's remaining sheep.

37. And Jacob took a rod of fresh poplar and one of almond-tree and one of plane-tree, and he peeled white stripes in them, exposing the white that was on the rods. 38. And he set up the rods that he had peeled in the channels in the watering troughs at which the flock came to drink, facing the flock. And they copulated when they came to drink. 39. And the flock copulated at the rods, and the flock gave birth to streaked, speckled, and spotted ones. 40. And Jacob separated the sheep and had the flock face the streaked and every brown one among Laban's flock, and he set his own droves apart and did not set them by Laban's flock. 41. And it was that whenever the fittest sheep would copulate Jacob would put the

rods in the channels before the sheep's eyes so that they would copulate by the rods, 42. and when the sheep were feebler he would not put them, so the feeble ones became Laban's and the fitter ones became Jacob's. 43. And the man expanded very, very much, and he had many sheep and female and male servants and camels and asses.

31:3. And YHWH said to Jacob, "Go back to your fathers' land and to your birthplace, and I'll be with you." 17. And Jacob got up and carried his children and his wives on the camels. 32:4. And Jacob sent messengers ahead of him to Esau, his brother, to the land of Seir, the territory of Edom, 5. and commanded them saying, "You shall say this: 'To my lord, to Esau, your servant Jacob said this, "I've resided with Laban and delayed until now, 6. and ox and ass and sheep and male and female servant have become mine. And I'm sending to tell my lord so as to find favor in your eyes."'"

7. And the messengers came back to Jacob saying, "We came to your brother, to Esau, and also he's coming to you—and four hundred men with him!" 8. And Jacob was very afraid, and he had anguish. And he divided the people who were with him and the sheep and the oxen and the camels into two camps. 9. And he said, "If Esau will come to one camp and strike it, then the camp that is left will survive."

10. And Jacob said, "God of my father Abraham and God of my father Isaac, YHWH, who said to me, 'Go back to your land and to your birthplace, and I'll deal well with you,' 11. I'm not worthy of all the kindnesses and all the faithfulness that you've done with your servant, because I crossed this Jordan with just my rod, and now I've become two camps. 12. Save me from my brother's hand, from Esau's hand, because I fear him, in case he'll come and strike me, mother with children. 13. And *you've* said, 'I'll do *well* with you, and I'll make your seed like the sand of the sea, that it won't be countable because of its great number.'"

33:1a. And Jacob raised his eyes and looked, and here was Esau coming, and four hundred men with him. 3b. And he bowed to the ground seven times until he came up to his brother. 4. And Esau ran to him and embraced him and fell on his neck and kissed him. And they wept. 16. And in that day Esau went on his way back to Seir.

34:1. And Dinah, Leah's daughter, whom she had borne to Jacob, went out to see the daughters of the land. 2. And Shechem, son of Hamor, the Hivite, the chieftain of the land, saw her. And he took her and lay with her and degraded her. 3. And his soul clung to Dinah, Jacob's daughter, and he loved the girl and spoke on the girl's heart. 4. And Shechem said to his father, Hamor, "Get me this girl for a wife."

5. And Jacob heard that he had defiled Dinah, his daughter, and his sons were with his cattle in the field, and Jacob kept quiet until they came. 6. And Hamor, Shechem's father, went out to Jacob to speak with him. 7. And Jacob's sons came from the field when they heard, and the people were pained, and they were very furious, for he had done a foolhardy thing among Israel, to lie with Jacob's daughter, and such a thing is not done. 8. And Hamor spoke with them, saying, "Shechem—my son—his soul longs for your daughter. Give her to him as a wife, 9. and marry with us; give your daughters to us and take our daughters to you, 10. and live with us; and the land will be before you: live, go around in it, and take possession in it."

11. And Shechem said to her father and to her brothers, "Let me find favor in your eyes, and I'll give whatever you say. 12. Make a bride price and gift on me very great, and let me give whatever you say to me, and give me the girl as a wife."

13. And Jacob's sons answered Shechem and his father Hamor with deception as they spoke because he had defiled their sister Dinah. 14. And they said to them, "We aren't able to do this thing, to give our sister to a man who has a foreskin, because that's a disgrace to us. 15. Only this way will we consent to you: if you'll be like us, every male among you to be circumcised. 16. And we'll give our daughters to you and take your daughters to us, and we'll live with you, and we'll become one people. 17. And if you won't listen to us, to be circumcised, then we'll take our daughter and go."

18. And their words were good in Hamor's eyes and in Hamor's son Shechem's eyes. 19. And the boy did not delay to do the thing, for he desired Jacob's daughter. And he was more respected than all his father's house. 20. And Hamor and his son Shechem came to the gate of their city and spoke to the people of their city, saying, 21. "These people are peaceable with us, and they'll live in the land and go around in it, and the land, here, has enough breadth for them. Let's take their daughters for us as wives and give our daughters to them. 22. Only in this way will these people consent to live with us, to be one people: if every male among us is circumcised as they are circumcised. 23. Their cattle and possessions and all their animals: won't they be ours? Only let's consent to them, and they'll live with us." 24. And everyone who went out of the gate of his city listened to Hamor and to his son Shechem; and every male, everyone who went out of the gate of his city, was circumcised.

25. And it was on the third day, when they were hurting, and two of Jacob's sons, Simeon and Levi, Dinah's brothers, each took his sword, and they came upon the city stealthily, and they killed every male. 26. And they killed Hamor and his son Shechem by the sword and took Dinah from Shechem's house and went out. 27. Jacob's sons had come upon the corpses and despoiled the city because they had defiled their sister. 28. They took their sheep and their oxen and their

asses and what was in the city and what was in the field. 29. And they captured and despoiled all their wealth and all their infants and their wives and everything that was in the house. 30. And Jacob said to Simeon and to Levi, "You've caused me anguish, making me odious to those who live in the land, to the Canaanite and to the Perizzite, and I'm few in number, and they'll be gathered against me and strike me, and I'll be destroyed, I and my house."

31. And they said, "Shall he treat our sister like a prostitute?"

35:21. And Israel traveled and pitched his tent past Migdal-Eder. 22. And it was when Israel was tenting in that land: and Reuben went and lay with Bilhah, his father's concubine. And Israel heard.

37:2b. Joseph, at seventeen years old, had been tending the sheep with his brothers, and he was a boy with the sons of Bilhah and the sons of Zilpah, his father's wives; and Joseph brought a bad report of them to their father. 3b. And he made him a coat of many colors.

5. And Joseph had a dream and told it to his brothers, and they went on to hate him more. 6. And he said to them, "Listen to this dream that I had: 7. and here we were binding sheaves in the field; and here was my sheaf rising and standing up, too; and, here, your sheaves surrounded and bowed to my sheaf."

8. And his brothers said to him, "Will you *rule* over us?! Will you *dominate* us?!" And they went on to hate him more because of his dreams and because of his words.

9. And he had yet another dream and told it to his brothers. And he said, "Here, I've had another dream, and here were the sun and the moon and eleven stars bowing to me." 10. And he told it to his father and to his brothers.

And his father was annoyed at him and said to him, "What is this dream that you've had? Shall *we* come, I and your mother and your brothers, to bow to you to the ground?!" 11. And his brothers were jealous of him, and his father took note of the thing.

19. And the brothers said to one another, "Here comes the dream-master, that one there! 20. And now, come on and let's kill him and throw him in one of the pits, and we'll say a wild animal ate him, and we'll see what his dreams will be!" 23. And it was when Joseph came to his brothers: and they took off Joseph's coat, the coat of many colors, which he had on.

25b. And they raised their eyes and saw, and here was a caravan of Ishmaelites coming from Gilead, and their camels were carrying spices and balsam and myrrh, going to bring them down to Egypt. 26. And Judah said to his brothers, "What profit is there if we kill our brother and cover his blood? 27. Come on and let's sell him to the Ishmaelites, and let our hand not be on him, because he's our brother, our flesh." And his brothers listened, 28b. and they sold Joseph to the Ishmaelites for twenty weights of silver. And they brought Joseph to Egypt.

31. And they took Joseph's coat and slaughtered a he-goat and dipped the coat in the blood. 32. And they sent the coat of many colors and brought it to their father and said, "We found this. Recognize: is it your son's coat or not?"

33. And he recognized it and said, "My son's coat. A wild animal ate him. Joseph is *torn up!*" 34. And Jacob ripped his clothes and wore sackcloth on his hips and mourned over his son many days. 35. And all his sons and his daughters got up to console him, and he refused to be consoled, and he said, "Because I'll go down mourning to my son at Sheol," and his father wept for him.

38:1. And it was at that time, and Judah went down from his brothers and turned to an Adullamite man, and his name was Hirah. 2. And Judah saw a daughter of a Canaanite man there, and his name was Shua. And he took her and came to her. 3. And she became pregnant and gave birth to a son, and he called his name Er. 4. And she became pregnant again and gave birth to a son, and she called his name Onan. 5. And she proceeded again to give birth to a son, and she called his name Shelah. (And he was at Chezib when she gave birth to him.) 6. And Judah took a wife for Er, his firstborn, and her name was Tamar. 7. And Er, Judah's firstborn, was bad in YHWH's eyes, and YHWH killed him. 8. And Judah said to Onan, "Come to your brother's wife and couple as a brother-in-law with her and raise seed for your brother." 9. And Onan knew that the seed would not be his. And it was when he came to his brother's wife: and he spent on the ground so as not to give seed for his brother. 10. And what he did was bad in YHWH's eyes, and He killed him, too.

11. And Judah said to Tamar, his daughter-in-law, "Live as a widow at your father's house until my son Shelah grows up" (because he said, "Or else he, too, will die like his brothers"). And Tamar went and lived at her father's house.

12. And the days were many, and Judah's wife, the daughter of Shua, died. And Judah was consoled. And he went up to his sheep-shearers, he and his friend Hirah, the Adullamite, to Timnah.

13. And it was told to Tamar, saying, "Here, your father-in-law is going up to Timnah to shear his sheep." 14. And she took off her widowhood clothes from on her and covered herself with a veil and wrapped herself and sat in a visible place that was on the road to Timnah, because she saw that Shelah had grown up and she had not been given to him as a wife. 15. And Judah saw her and thought her to be a prostitute because she had covered her face. 16. And he turned to her by the road.

And he said, "Come on. Let me come to you," because he did not know that she was his daughter-in-law.

And she said, "What will you give me when you come to me?"

17. And he said, "I'll have a goat kid sent from the flock."

And she said, "If you'll give a pledge until you send it."

18. And he said, "What is the pledge that I'll give you?"

And she said, "Your seal and your cord and your staff that's in your hand." And he gave them to her, and he came to her, and she became pregnant by him.

19. And she got up and went and took off her veil from on her and put on her widowhood clothes. 20. And Judah sent the goat kid by the hand of his friend the Adullamite to take back the pledge from the woman's hand, and he did not find her. 21. And he asked the people of her place, saying, "Where is the sacred prostitute? She was visibly by the road."

And they said, "There was no sacred prostitute here."

22. And he went back to Judah and said, "I didn't find her, and also the people of the place said 'There was no sacred prostitute here.'"

23. And Judah said, "Let her take them or else we'll be a disgrace. Here, I've sent this goat kid, and you didn't find her."

24. And it was about three months, and it was told to Judah, saying, "Your daughter-in-law Tamar has whored, and, here, she's pregnant by whoring as well."

And Judah said, "Bring her out and let her be burned."

25. She was brought out. And she sent to her father-in-law, saying, "I'm pregnant by the man to whom these belong," and she said, "Recognize: to whom do these seal and cords and staff belong?"

26. And Judah recognized and said, "She's more right than I am, because of the fact that I didn't give her to my son Shelah." And he did not go on to know her again.

27. And it was at the time that she was giving birth, and here were twins in her womb. 28. And it was as she was giving birth, and one put out his hand, and the midwife took a scarlet thread and tied it on his hand, saying, "This one came out first." 29. And it was as he pulled his hand back, and here his brother came out. And she said, "What a breach you've made for yourself!" And he called his name Perez. 30. And his brother who had the scarlet thread on his hand came out after. And he called his name Zerah.

39:1. And Joseph had been brought down to Egypt. And an Egyptian man, Potiphar, an official of Pharaoh, chief of the guards, bought him from the hand of the Ishmaelites who had brought him down there. 2. And YHWH was with Joseph, and he was a successful man, and he was in his Egyptian lord's house. 3. And his lord saw that YHWH was with him, and YHWH made everything he did successful in his hand. 4. And Joseph found favor in his eyes, and he attended him, and he appointed him over his house and put everything he had in his hand. 5. And it was from the time that he appointed him in his house and over all that he had that YHWH blessed the Egyptian's house on account of Joseph, and YHWH's blessing was in everything that he had, in the house and in the field. 6. And he left everything that he had in Joseph's hand and did not know a thing about what he had except the bread that he was eating.

And Joseph had an attractive figure and was handsome. 7. And it was after these things, and his lord's wife raised her eyes to Joseph and said, "Lie with me."

8. And he refused, and he said to his lord's wife, "Here, my lord doesn't know what he has with me in the house, and he's put everything that he has in my hand. 9. No one is bigger than I am in

this house, and he hasn't held back a thing from me except you because you're his wife, and how could I do this great wrong? And I would sin against God." 10. And it was, when she spoke to Joseph day after day, that he didn't listen to her, to lie by her, to be with her. 11. And it was on a day like this, and he came to the house to do his work, and not one of the people of the house was there in the house.

12. And she caught him by his garment, saying, "Lie with me!" And he left his garment in her hand and fled and went outside. 13. And it was, when she saw that he had left his garment in her hand and run off outside, 14. that she called to the people of her house and said to them, saying, "See, he brought us a Hebrew man to fool with us. He came to me, to lie with me, and I called in a big voice. 15. And it was when he heard that I raised my voice and called that he left his garment by me and fled and went outside." 16. And she lay his garment down by her until his lord came to his house. 17. And she spoke things like these to him, saying, "The Hebrew slave, which you brought us, came to me, to fool with me. 18. And it was when I raised my voice and called that he left his garment by me and fled outside." 19. And it was, when his lord heard his wife's words that she spoke to him, saying, "Your servant did things like these to me," that his anger flared. 20. And Joseph's lord took him and put him in prison, a place where the king's prisoners were kept.

And he was there in the prison, 21. and YHWH was with Joseph and extended kindness to him and gave him favor in the eyes of the warden. 22. And the warden put all the prisoners who were in the prison into Joseph's hand, and he was doing all the things that they do there. 23. The warden was not seeing anything in his hand because YHWH was with him, and YHWH would make whatever he did successful.

[No story of Joseph's rise from prison to high rank is preserved.]

42:1. And Jacob saw that there was grain in Egypt, and Jacob said to his sons, "Why do you look at each other?" 2. And he said, "Here I've heard that there's grain in Egypt. Go down there and buy grain for us from there, and we'll live and not die." 3. And ten of Joseph's brothers went down to buy grain from Egypt, 4. and Jacob did not send Joseph's brother Benjamin with his brothers because, he said, "In case some harm will happen to him."

6. And Joseph was the one in charge over the land; he was the one who sold grain to all the people of the land. And Joseph's brothers came and bowed to him, noses to the ground. 8. And Joseph recognized his brothers, and they did not recognize him, 9. and Joseph remembered the dreams that he had had about them. And he said to them, "You're spies. You came to see the land exposed."

10. And they said to him, "No, my lord, your servants came to buy food. 11. We're all sons of one man. We're honest men. Your servants weren't spying."

12. And he said to them, "No, but you've come to see the land exposed."

13. And they said, "Your servants are twelve brothers. We're sons of one man in the land of Canaan, and, here, the youngest is with our father today, and one is no more."

14. And Joseph said to them, "It's as I spoke to you, saying: you're spies. 15. By this you'll be tested: by Pharaoh's life, you won't go out of here except by your youngest brother's coming here. 16. Send one from among you, and let him get your brother,

and you'll be kept in prison, and your words will be tested: is the truth with you. And if not, then by Pharaoh's life you're spies." 17. And he gathered them under watch for three days. 18. And Joseph said to them on the third day, "Do this and live, as I fear God: 19. If you're honest, one brother from among you will be held at the place where you're under watch; and you, go, bring grain for the famine in your houses. 20. And you will bring your youngest brother to me, and your words will be confirmed, and you won't die." And they did so. 26. And they loaded their grain on their asses and went from there.

27. And one opened his sack to give fodder to his ass at a lodging place, and he saw his silver, and here it was in the mouth of his bag. 28. And he said to his brothers, "My silver's been put back, and here it is in my bag, too."

And their heart went out, and the brothers trembled to one another, saying, "What is this that God has done to us?"

29. And they came to Jacob, their father, to the land of Canaan, and told him all the things that happened to them, saying, 30. "The man, the lord of the land, spoke with us hard and accused us of spying on the land. 31. And we said to him, 'We're honest. We weren't spying. 32. We're twelve brothers, our father's sons, one is no more, and the youngest is with our father in the land of Canaan today.' 33. And the man, the lord of the land, said to us, 'By this I'll know that you're honest: leave one brother from among you with me, and take for the famine in your houses and go, 34. and bring your youngest brother to me, so I may know that you aren't spies, that you're honest. I'll give you your brother, and you'll go around in the land.'"

38. And he said, "My son will not go down with you. Because his brother's dead, and he's left by himself, and if some harm would happen to him on the way in which you're going then you'll bring down my gray hair in anguish to Sheol."

43:1. And the famine was heavy in the land. 2. And it was when they had finished eating the grain that they had brought from Egypt: and their father said to them, "Go back. Buy us a little food."

3. And Judah said to him, saying, "The man *certified* to us, saying, 'You won't see my face unless your brother is with you.' 4. If you're sending our brother with us, we'll go down and buy food for you, 5. and if you're not sending, we won't go down, because the man said to us, 'You won't see my face unless your brother is with you.'"

6. And Israel said, "Why have you done me wrong, to tell the man that you have another brother?"

7. And they said, "The man *asked* about us and about our birthplace, saying, 'Is your father still alive? Do you have a brother?' And we told him about these things. Could we have *known* that he would say, 'Bring your brother down'?"

8. And Judah said to Israel, his father, "Send the boy with me, so we may get up and go, and we'll live and not die, we and you and our infants as well. 9. I'll be security for him. You'll seek him from my hand. If I don't bring him to you and set him before you then I'll have sinned against you for all time. 10. For, if we hadn't delayed, by now we would have come back twice."

11. And Israel, their father, said to them, "If that's how it is, then do this: Take some of the best fruit of the land in your containers and take a gift down to the man, a little balm and a little honey, gum and myrrh, pistachios and almonds. 12. And take double the silver in your hand, and take back in your hand the silver that was put back in the mouth of your bags. Maybe it was a mistake. 13. And take your brother. And get up, go back to the man."

15. And the men took this gift and took double the silver in their hand and Benjamin and got up and went down to Egypt and stood

before Joseph. 16. And Joseph saw Benjamin with them and said to the one who was over his house, "Bring the men to the house, and slaughter and prepare an animal, because the men will eat with me at noon." 17. And the man did as Joseph said, and the man brought the men to Joseph's house.

18. And the men were afraid because they were brought to Joseph's house, and they said, "We're being brought on account of the silver that came back in our bags the first time, in order to roll over us and to fall upon us and to take us as slaves—and our asses." 19. And they went over to the man who was over Joseph's house and spoke to him at the entrance of the house 20. and said, "Please, my lord, we *came down* the first time to buy food, 21. and it was when we came to the lodging place, and we opened our bags, and here was each man's silver in the mouth of his bag, our money in its full weight. And we've brought it in our hand, 22. and we've brought down additional silver in our hand to buy food. We don't know who put our silver in our bags."

23a. And he said, "Peace to you. Don't be afraid. Your God and your father's God put treasure in your bags for you. Your silver came to me." 24. And the man brought the men to Joseph's house, and he gave water, and they washed their feet, and he gave fodder for their asses. 25. And they prepared the gift until Joseph's arrival at noon because they heard that they would eat bread there. 26. And Joseph came to the house, and they brought the gift for him that was in their hand to the house, and they bowed to him to the ground.

27. And he asked if they were well, and he said, "Is your old father whom you mentioned well? Is he still alive?"

28. And they said, "Your servant, our father, is well. He's still alive."

And he said, "Blessed of God is that man." And they knelt and bowed.

29. And he raised his eyes and saw Benjamin, his brother, his mother's son, and said, "Is this your youngest brother whom you mentioned to me?" And he said, "May God be gracious to you, my son." 30. And Joseph hurried because his feelings for his brother were boiling, and he sought a place to weep and came to his room and wept there. 31. And he washed his face and went out and restrained himself and said, "Put out bread." 32. And they put it out for him by himself and for them by themselves and for the Egyptians who were eating with him by themselves, because the Egyptians could not eat bread with the Hebrews because that is an appalling thing to Egypt. 33. And they sat before him, the first-born according to his birthright and the youngest according to his youth. And the men looked amazed at one another. 34. And he conveyed portions from before him to them, and he made Benjamin's portion five times more than the portions of all of them, and they drank and were drunk with him.

44:1. And he commanded the one who was over his house, saying, "Fill the men's bags with as much food as they can carry and put each man's silver in the mouth of his bag, 2. and put my cup, the silver cup, in the mouth of the youngest one's bag, and the silver for his grain." And he did according to Joseph's word that he spoke.

3. The morning was light: and the men had been sent away, they and their asses. 4. They had gone out of the city. They had not gone far, and Joseph had said to the one who was over his house, "Get up. Pursue the men, and you'll overtake them and say to them, 'Why did you pay back bad for good? 5. Isn't this the thing from which my lord drinks? And he *divines* by it! You've done bad, this thing that you've done.'" 6. And he overtook them and spoke these things to them.

7. And they said to him, "Why would my lord speak things like these? Far be it from your servants to do a thing like this. 8. Here,

we brought the silver that we found in the mouth of our bags back to you from the land of Canaan, so how would we steal silver or gold from your lord's house?! 9. The one among your servants with whom it's found, let him die, and also we'll become my lord's servants."

10. And he said, "Now, also, it will be so according to your words: the one with whom it's found will become my servant. And you will be innocent."

11. And they hurried, and each man lowered his bag to the ground, and each man opened his bag. 12. And he searched. With the oldest he began, and with the youngest he finished. And the cup was found in Benjamin's bag. 13. And they ripped their clothes, and each man loaded his ass, and they went back to the city. 14. And Judah and his brothers came to Joseph's house, and he was still there, and they fell to the ground before him.

15. And Joseph said to them, "What is this thing that you've done? Didn't you know that a man like me would *divine?"*

16. And Judah said, "What shall we say to my lord? What shall we speak? By what shall we justify ourselves? God has found your servants' crime. Here we're my lord's servants, both we and the one in whose hand the cup was found."

17. And he said, "Far be it from me to do this. The man in whose hand the cup was found: he will be my servant; and you, go up in peace to your father."

18. And Judah went over to him and said, "Please, my lord, let your servant speak something in my lord's ears, and let your anger not flare at your servant, because you're like Pharaoh himself. 19. My lord asked his servants, saying, 'Do you have a father or brother?' 20. and we said to my lord, 'We have an old father and a young son of his old age, and his brother's dead, and he's left alone of his mother, and his father loves him.' 21. And you said to your

servants, 'Bring him down to me, so I may set my eye on him.' 22. And we said to my lord, 'The boy can't leave his father; if he left his father he'd die.' 23. And you said to your servants, 'If your youngest brother doesn't come down with you, you won't see my face again.' 24. And it was, when we went up to your servant, my father, and told him my lord's words, 25. that our father said, 'Go back. Buy us a little food.' 26. And we said, 'We can't go down. If our youngest brother is with us then we'll go down, because we can't see the man's face if our youngest brother isn't with us.' 27. And your servant, my father, said to us, 'You know that my wife gave birth to two for me, 28. and one went away from me, and I said he's surely *torn up,* and I haven't seen him since. 29. And if you take this one from me as well and some harm happens to him, then you'll bring down my gray hair in wretchedness to Sheol.' 30. And now, when I come to your servant, my father, and the boy isn't with us, and he's bound to him soul to soul, 31. it will be, when he sees that the boy isn't there, that he'll die. And your servants will have brought down your servant our father's gray hair in anguish to Sheol, 32. because your servant offered security for the boy to my father, saying, 'If I don't bring him to you then I'll have sinned against my father for all time.' 33. And now, let your servant stay as my lord's servant in place of the boy, and let the boy go up with his brothers—34. for how could I go up to my father and the boy isn't with me—or else I'll see the wretchedness that will find my father."

45:1. And Joseph was not able to restrain himself in front of everyone who was standing by him, and he called, "Take everyone out from my presence." And not a man stood with him when Joseph made himself known to his brothers. 2. And he wept out loud. And Egypt heard, and Pharaoh's house heard. 4. And Joseph said to his brothers, "Come over to me." And they went over. And he said, "I'm Joseph, your brother, whom you sold to Egypt. 5. And now, don't be sad and let there be no anger in your eyes because you sold me here, because God sent me before you to preserve life. 6. For it's two years that the famine is in the land, and for five more years there'll be no plowing and harvest. 7. And God sent me ahead of you to provide a remnant for you in the earth and to keep you alive as a big, surviving community. 8. And now, it wasn't you who sent me here, but God, and he made me into a father to Pharaoh and a lord to all his house and a ruler in all the land of Egypt. 9. Hurry, and go up to my father and say to him: 'Your son Joseph said this: God has made me a lord to all Egypt. Come down to me. Don't stand back. 10. And you'll live in the land of Goshen, and you'll be close to me, you and your children and your children's children and your flock and your oxen and everything you have. 11. And I'll provide for you there, for there are five more years of famine, or else you and your house and everything you have will come to poverty.' 12. And here your eyes see, and the eyes of my brother Benjamin, that it's my mouth that speaks to you. 13. And you'll tell my father of all my glory in Egypt and of all that you've seen, and hurry and bring my father down here." 14. And he fell on his brother Benjamin's neck and wept, and Benjamin wept on his neck. 15. And he kissed all his brothers and wept over them. And after that his brothers spoke with him.

16. And the report was heard at Pharaoh's house, saying, "Joseph's brothers have come." And it was good in Pharaoh's eyes and in his servants' eyes. 17. And Pharaoh said to Joseph, "Say to your brothers, 'Do this: load your beasts and go; come to the land of Canaan, 18. and take your father and your households and come to me, and I'll give you the best of the land of Egypt; and eat the fat of the land.' 19. And you are commanded, 'Do this: Take wagons from

the land of Egypt for your infants and for your wives, and carry your father and come. 20. And let your eye not care about your possessions, for the best of all the land of Egypt is yours.'"

21. And the children of Israel did so. And Joseph gave them wagons by order of Pharaoh and gave them provisions for the road. 22. For all of them he gave each man changes of clothes, and he gave Benjamin three hundred weights of silver and five changes of clothes. 23. And to his father he sent as follows: ten asses bearing Egypt's best, and ten she-asses bearing grain, bread, and food supply for his father for the road. 24. And he sent his brothers off, and they went, and he said to them, "Don't quarrel on the road."

25. And they went up from Egypt and came to the land of Canaan, to Jacob, their father, 26. and told him, saying, "Joseph is still alive! And he rules all the land of Egypt!" And his heart grew numb because he did not believe them. 27. And they spoke to him all of Joseph's words that he had spoken to them, and he saw the wagons that Joseph had sent to carry him, and their father Jacob's spirit lived.

28. And Israel said, "Enough. Joseph, my son, is still alive. Let me go and see him before I die."

46:5b. And the children of Israel carried Jacob, their father, and their infants and their wives in the wagons that Pharaoh had sent to carry him. 28. And he sent Judah ahead of him to Joseph to direct him in advance to Goshen. And they came to the land of Goshen. 29. And Joseph hitched his chariot and went up to Israel, his father, at Goshen. And he appeared to him and fell on his neck and wept on his neck a long time. 30. And Israel said to Joseph, "Let me die now after I've seen your face, that you're still alive."

31. And Joseph said to his brothers and to his father's house, "Let me go up and tell Pharaoh and say to him, 'My brothers and my father's house that were in the land of Canaan have come to

me. 32. And the people are shepherds, because they had been cattle people, and they've brought their flock and their oxen and all they have.' 33. And it will be that Pharaoh will call you and say, 'What is your occupation?' 34. And you'll say, 'Your servants were cattle people from our youth until now, both we and our fathers,' so that you'll live in the land of Goshen, because any shepherd is an appalling thing to Egypt." 47:1. And Joseph came and told Pharaoh and said, "My father and my brothers and their flock and their oxen and all that they own have come from the land of Canaan, and here they are in the land of Goshen." 2. And he took several of his brothers, five men, and set them before Pharaoh.

3. And Pharaoh said to his brothers, "What is your work?"

And they said to Pharaoh, "Your servants are shepherds, both we and our fathers," 4. and they said to Pharaoh, "We came to reside in the land because there's no pasture for your servants' flock because the famine is heavy in the land of Canaan. And now may your servants live in the land of Goshen."

5. And Pharaoh said to Joseph, saying, "Your father and your brothers have come to you. 6. The land of Egypt is before you. Settle your father and your brothers in the best of the land. Let them live in the land of Goshen. And if you know—and if there are among them—worthy men, then you shall make them cattle officers over those that I have." 27a. And Israel lived in Egypt in the land of Goshen, and they held property in it.

29. And Israel's days to die drew close. And he called his son, Joseph, and said to him, "If I've found favor in your eyes, place your hand under my thigh and practice kindness and faithfulness with me: don't bury me in Egypt. 30. And I'll lie with my fathers, and you'll carry me out of Egypt and bury me in their burial place."

And he said, "I'll do according to your word."

31. And he said, "Swear to me."

And he swore to him. And Israel bowed at the head of the bed.

49:1. And Jacob called to his sons and said:

Gather, and I'll tell you what will happen to you in the
future days.

2. Assemble and listen, sons of Jacob,
and listen to Israel, your father.

3. Reuben, you're my firstborn,
my power, and the beginning of my might,
preeminent in bearing and preeminent in strength.

4. Unstable as water, you'll not be preeminent,
for you ascended your father's bed;
then you defiled, going up to my couch.

5. Simeon and Levi are brothers:
Implements of violence are their tools of trade.

6. Let my soul not come in their council;
let my glory not be united in their society.
For in their anger they killed a man,
and by their will they crippled an ox.

7 Cursed is their anger, for it's strong,
and their wrath, for it's hard.
I'll divide them in Jacob,
and I'll scatter them in Israel.

8. Judah: You, your brothers will praise you.
Your hand on your enemies' neck,
your father's sons will bow to you.

9. A lion's whelp is Judah;
from prey, my son, you've risen.
He bent, crouched, like a lion;

and, like a lioness, who will rouse him?
10. The scepter won't depart from Judah
or a ruler from between his legs
until he comes to Shiloh,
and peoples' obedience is his.
11. Tying his ass to the vine
and his she-ass's foal to the choice vine,
he washed in wine his clothing
and in blood of grapes his garment,
12. eyes darker than wine
and teeth whiter than milk.
13. Zebulon will settle by seashores:
and he'll be a shore for boats,
and his border at Sidon.
14. Issachar is a strong ass
crouching between the saddle-packs:
15. and he saw rest, that it was good,
and the land, that it was pleasant,
and he leaned his shoulder to bear
and became a corvée slave.
16. Dan will judge his people
as one of the tribes of Israel:
17. Dan will be a snake on a road,
a venomous snake on a path,
that bites a horse's heels,
and its rider falls backward.

18. I wait for your salvation, YHWH.

19. Gad: a troop will trap him,
    and he'll trap their heel.
20. From Asher: his bread will be rich,
    and he'll provide a king's delights.
21. Naphtali: a hind let loose,
    who gives lovely words.
22. A fruitful bough is Joseph,
    a fruitful bough over a pool,
    branches running over a wall:
23. And archers bitterly attacked him,
    shot at him, and despised him.
24. And his bow stayed strong,
    and his forearms were nimble,
    from the hands of the Mighty One of Jacob,
    from there, the shepherd, the rock of Israel,
25. from your father's God, and He'll strengthen you,
    and Shadday, and He'll bless you,
    blessings of skies from above,
    blessings of deep, crouching below,
    blessings of breast and womb.
26. blessings of your father, the mighty and most high,
    blessings of the mountains of old
    desired object of the hills of antiquity.
    They'll be on Joseph's head,
    on the top of the head of the one separate from his
      brothers.
27. Benjamin is a tearing wolf:
    in the morning eating prey,
    and at evening dividing booty.

[A report of Jacob's death is not present.]

50:1. And Joseph fell upon his father's face and wept over him and kissed him. 2. And Joseph commanded his servants, the physicians, to embalm his father. And the physicians embalmed Israel, 3. and they spent forty days on him, because that is the number of days they spent on the embalming, and Egypt mourned him seventy days. 4. And his mourning days passed, and Joseph spoke to Pharaoh's house, saying, "If I've found favor in your eyes, speak in Pharaoh's ears, saying, 5. 'My father had me swear, saying, "Here I'm dying. In my tomb, which I dug for me in the land of Canaan: you shall bury me there." And now, let me go up and bury my father and come back.'"

6. And Pharaoh said, "Go up and bury your father as he had you swear."

7. And Joseph went up to bury his father, and all of Pharaoh's servants, the elders of his house, and all the elders of the land of Egypt, 8. and all of Joseph's house and his brothers and his father's house. Only their infants and their flock and their oxen they left in the land of Goshen. 9. And both chariots and horsemen went up with him. It was a very heavy camp. 10. And they came to the threshing floor of Atad, which is across the Jordan, and they had a very big and heavy funeral, and he made a mourning for his father for seven days. 11. And the Canaanite residents of the land saw the mourning at the threshing floor of Atad and said, "This is a heavy mourning to Egypt." On account of this its name was called Abel of Egypt, which is beyond the Jordan. 14. And, after his burial of his father, Joseph went back to Egypt, he and his brothers and all those who went up with him to bury his father.

22b. And Joseph lived one hundred ten years. Exodus 1:6. And Joseph and all of his brothers and all of that generation died.

[The account of the oppression of the children of Israel is lost.]

22. And Pharaoh commanded all of his people, saying, "Every son who is born: you shall throw him into the Nile. And every daughter you shall keep alive."

2:1. And a man from the house of Levi went and took a daughter of Levi. 2. And the woman became pregnant and gave birth to a son. And she saw him, that he was good, and she concealed him for three months. 3. And she was not able to conceal him anymore, and she took an ark made of bulrushes for him and smeared it with bitumen and with pitch and put the boy in it and put it in the reeds by the shore of the Nile. 4. And his sister stood still at a distance to know what would be done to him. 5. And the Pharaoh's daughter went down to bathe at the Nile, and her girls were going alongside the Nile, and she saw the ark among the reeds and sent her maid, and she took it. 6. And she opened it and saw him, the child: and here was a boy crying, and she had compassion on him, and she said, "This is one of the Hebrews' children."

7. And his sister said to Pharaoh's daughter, "Shall I go and call a nursing woman from the Hebrews for you, and she'll nurse the child for you?"

8. And Pharaoh's daughter said to her, "Go." And the girl went and called the child's mother. 9. And Pharaoh's daughter said to her, "Take this child and nurse him for me, and I'll pay your wage." And the woman took the boy and nursed him. 10. And the boy grew older, and she brought him to Pharaoh's daughter, and he

became her son. And she called his name Moses, and she said, "Because I drew him from the water."

11. And it was in those days, and Moses grew older, and he went out to his brothers and saw their burdens, and he saw an Egyptian man striking a Hebrew man, one of his brothers. 12. And he turned this way and that way and saw that there was no man, and he struck the Egyptian and hid him in the sand. 13. And he went out on the second day, and here were two Hebrew men fighting. And he said to the one who was in the wrong, "Why do you strike your companion?"

14. And he said, "Who made you a commander and judge over us? Are you saying you'd kill me—the way you killed the Egyptian?!"

And Moses was afraid and said, "The thing is known for sure." 15. And Pharaoh heard this thing and sought to kill Moses, and Moses fled from Pharaoh's presence and lived in the land of Midian.

And he sat by a well. 16. And a priest of Midian had seven daughters, and they came and drew water and filled the troughs to water their father's flock, 17. and the shepherds came and drove them away. And Moses got up and saved them and watered their flock. 18. And they came to Reuel, their father, and he said, "Why were you so quick to come today?"

19. And they said, "An Egyptian man rescued us from the shepherds' hand, and he *drew water* for us and watered the flock, too."

20. And he said to his daughters, "And where is he? Why is this that you've left the man? Call him, and let him eat bread."

21. And Moses was content to live with the man. And he gave Zipporah, his daughter, to Moses, 22. and she gave birth to a son, and he called his name Gershom, "Because," he said, "I was an alien in a foreign land."

23a. And it was after those many days, and the king of Egypt died.

3:2. And an angel of YHWH appeared to him in a fire's flame

from inside a bush. And he looked, and here was the bush burning in the fire, and the bush was not consumed! 3. And Moses said, "Let me turn and see this great sight. Why doesn't the bush burn?"

4a. And YHWH saw that he turned to see, 5. and He said, "Don't come close here. Take off your shoes from your feet, because the place on which you're standing: it's holy ground." 7. And YHWH said, "I've *seen* the degradation of my people who are in Egypt, and I've heard their wail on account of their oppressors, because I know their pains. 8. And I've come down to rescue them from Egypt's hand and to bring them up from that land to a good and widespread land, to a land flowing with milk and honey, to the place of the Canaanite and the Hittite and the Amorite and the Perizzite and the Hivite and the Jebusite." 4:19. And YHWH said to Moses in Midian, "Go. Go back to Egypt, because all the people who were seeking your life have died."

20a. And Moses took his wife and his son and rode them on an ass, and he went back to the land of Egypt. 24. And he was on the way, at a lodging place, and YHWH met him, and he asked to kill him. 25. And Zipporah took a flint and cut her son's foreskin and touched his feet, and she said, "Because you're a bridegroom of blood to me." 26. And he held back from Him. Then she said, "A bridegroom of blood for circumcisions."

5:1. And after that Moses and Aaron came and said to Pharaoh, "YHWH, God of Israel, said this: 'Let my people go, so they may celebrate a festival for me in the wilderness.'"

2. And Pharaoh said, "Who is YHWH that I should listen to His voice, to let Israel go?! I don't know YHWH, and also I won't let Israel go."

[No story of plagues or of Israel's departure is present.]

13:21. And YHWH was going in front of them by day in a column of cloud to show them the way, and by night in a column of fire to shed light for them, so as to go by day and by night. 22. The column of cloud by day and the column of fire by night did not depart in front of the people.

14:5a. And it was told to the king of Egypt that the people had fled. 6. And he hitched his chariot and took his people with him, 9a. and Egypt pursued them. 10b. And the children of Israel raised their eyes, and here was Egypt coming after them, and they were very afraid. 13. And Moses said to the people, "Don't be afraid. Stand still and see YHWH's salvation that He'll do for you today. For, as you've seen Egypt today, you'll never see them again, ever. 14. YHWH will fight for you, and you'll keep quiet!" 19b. And the column of cloud went from in front of them and stood behind them. 20b. And there was the cloud and darkness [for the Egyptians], while it [the column of fire] lit the night [for the Israelites], and one did not come near the other all night. 21b. And YHWH drove back the sea with a strong east wind all night and turned the sea into dry ground. 24. And it was in the morning watch, and YHWH gazed at Egypt's camp in a column of fire and cloud and threw Egypt's camp into tumult.

25b. And Egypt said, "Let me flee from Israel, because YHWH is fighting for them against Egypt." 27b. And the sea had gone back to its natural level toward morning, and Egypt was fleeing toward it. And YHWH tossed the Egyptians into the sea. 30. And YHWH saved Israel from Egypt's hand that day. And Israel saw Egypt dead on the seashore, 31. and Israel saw the big hand that YHWH had used against Egypt, and the people feared YHWH, and they trusted in YHWH and in Moses His servant.

16:4. And YHWH said to Moses, "Here, I'm raining bread from the skies for you, and let the people go out and gather the daily ration in its day in order that I can test them: will they go by my instruction or not. 5. And it will be that on the sixth day they'll prepare what they bring in, and it will be twice as much as they'll gather daily." 35b. They ate the manna until they came to the edge of the land of Canaan.

[No report of the arrival at Sinai is present.]

19:10. And YHWH said to Moses, "Go to the people and consecrate them today and tomorrow; and they shall wash their clothes 11. and be ready for the third day, because on the third day YHWH will come down on Mount Sinai before the eyes of all the people. 12. And you shall limit the people all around, saying, 'Watch yourselves about going up in the mountain and touching its edge. Anyone who touches the mountain will be *put to death.* 13. A hand shall not touch him, but he will be *stoned* or *shot.* Whether animal or man, he will not live.' At the blowing of the horn they shall go up the mountain."

14. And Moses went down from the mountain to the people. And he consecrated the people, and they washed their clothes. 15. And he said to the people, "Be ready for three days. Don't come close to a woman."

16a. And it was on the third day, when it was morning, 18. and Mount Sinai was all smoke because YHWH came down on it in fire, and its smoke went up like the smoke of a furnace, and the whole mountain rumbled greatly. 20. And YHWH came down on Mount Sinai, at the top of the mountain, and YHWH called to Moses at the top of the mountain, and Moses went up. 21. And

YHWH said to Moses, "Go down. Warn the people in case they break through to YHWH, to see, and many of them fall. 22. And also let the priests who approach YHWH consecrate themselves, or else YHWH will break out against them."

23. And Moses said to YHWH, "The people is not able to go up to Mount Sinai, because you warned us, saying, 'Limit the mountain and consecrate it.'"

24. And YHWH said to him, "Go. Go down. Then you'll come up. And let the priests and the people not break through to come up to YHWH, or else He'll break out against them." 25. And Moses went down to the people, and he said to them . . .

34:1a. And YHWH said to Moses, "Carve two tablets of stones, 2. and be ready for the morning, and you shall go up in the morning to Mount Sinai and present yourself to me there on the top of the mountain. 3. And no man shall go up with you, and also let no man be seen in all of the mountain. Also let the flock and the oxen not feed before that mountain." 4. And he carved two tablets of stones, and Moses got up early in the morning and went up to Mount Sinai as YHWH had commanded him, and he took in his hand two tablets of stones. 5. And YHWH came down in a cloud and stood with him there, and he invoked the name YHWH. 6. And YHWH passed in front of him and called,

> YHWH, YHWH, merciful and gracious God, slow to anger and abounding in kindness and faithfulness, 7. keeping kindness for thousands, bearing crime and offense and sin; though not making one innocent: reckoning fathers' crime on children and on children's children, on third generations and on fourth generations.

8. And Moses hurried and knelt to the ground and bowed, 9. and he said, "If I've found favor in your eyes, my Lord, may my Lord go among us, because it is a stiff-necked people, and forgive our crime and our sin, and make us your possession."

10. And He said, "Here, I'm making a covenant. Before all your people I'll do wonders that haven't been created in all the earth and among all the nations; and all the people whom you're among will see YHWH's deeds, because that which I'm doing with you is awesome. 11. Watch yourself regarding what I command you today. Here, I'm driving out from before you the Amorite and the Canaanite and the Hittite and the Perizzite and the Hivite and the Jebusite. 12. Be watchful that you don't make a covenant with the resident of the land onto which you're coming, that he doesn't become a trap among you. 13. But you shall demolish their altars and shatter their pillars and cut down their Asherahs.

14. For you shall not bow to another god—because YHWH: His name is Jealous, He is a jealous God—15. that you not make a covenant with the resident of the land, and they will prostitute themselves after their gods and sacrifice to their gods, and he will call to you, and you will eat from his sacrifice. 16. And you will take his daughters for your sons, and his daughters will prostitute themselves after their gods and cause your sons to prostitute themselves after their gods.

17. You shall not make molten gods for yourself.

18. You shall observe the Festival of Unleavened Bread. Seven days you shall eat unleavened bread, which I commanded you, at the appointed time, the month of Abib; because in the month of Abib you went out of Egypt.

19. Every first birth of a womb is mine, and all your animals that have a male first birth, ox or sheep. 20. And you shall redeem an ass's first birth with a sheep, and if you do not

redeem it then you shall break its neck. You shall redeem every firstborn of your sons. And none shall appear before me empty-handed.

21. Six days you shall work, and in the seventh day you shall cease. In plowing time and in harvest, you shall cease.

22. And you shall make yourselves a Festival of Weeks, of the firstfruits of the wheat harvest, and the Festival of Gathering at the end of the year. 23. Three times in the year every one of your males shall appear before the Lord YHWH, God of Israel. 24. For I shall dispossess nations before you and extend your border, and no man will covet your land while you are going up to appear before YHWH, your God, three times in the year.

25. You shall not offer the blood of my sacrifice on leavened bread.

And the sacrifice of the Festival of Passover shall not remain until the morning.

26. You shall bring the first of the firstfruits of your land to the house of YHWH, your God.

You shall not cook a kid in its mother's milk.

27. And YHWH said to Moses, "Write these things for yourself, because I have made a covenant with you and with Israel based upon these things." 28. And he was there with YHWH forty days and forty nights. He did not eat bread, and he did not drink water. And he wrote on the tablets the words of the covenant, the Ten Commandments.

Numbers 10:29. And Moses said to Hobab, the son of Reuel the Midianite, Moses' father-in-law, "We're traveling to the place that YHWH said, 'I'll give it to you.' Come with us, and we'll be good to you, because YHWH has spoken of good regarding Israel."

30. And he said to him, "I won't go, but rather I'll go to my land and to my birthplace."

31. And he said, "Don't leave us, because you know the way we should camp in the wilderness, and you'll be eyes for us. 32. And it will be, when you go with us, that we'll do good for you in proportion to the good that YHWH will do for us."

33. And they traveled from the mountain of YHWH three days' journey. And the ark of the covenant of YHWH traveled in front of them three days' journey to scout a resting place for them. 34. And YHWH's cloud was over them by day as they traveled from the camp. 35. And it was, when the ark traveled, that Moses said, "Arise, YHWH, and let your enemies be scattered, and let those who hate you flee from your presence," 36. and when it rested, he said, "Come back, YHWH, the ten thousands of thousands of Israel."

13:17. And Moses sent [people] to scout the land of Canaan. And he said to them, "Go up there in the Negeb and go up the mountain 18. and see the land, how it is; and the people who live on it, are they strong or weak, are they few or many; 19. and how is the land in which they live, is it good or bad; and how are the cities in which they live, are they in camps or in fortified places; 20. and how is the land, is it fat or meager; does it have trees or not; and exert strength and take some fruit of the land."

And it was the days of the first grapes. 22. And they went up in the Negeb and came to Hebron. And Ahiman, Sheshai, and Talmai, the offspring of the giants, were there. (And Hebron was built seven years before Zoan of Egypt.) 23. And they came to the wadi Eshcol, and they cut a branch and one cluster of grapes from there—and carried it on a pole by two people—and some pomegranates and some figs. 24. That place was called wadi Eshcol on account of the cluster that the children of Israel cut from there. 27. And they told [Moses] and said, "We came to the land where you sent us, and also it's flowing with milk and honey, and this is its fruit. 28. Nonetheless: the people who live in the land are strong. And the cities are fortified, very big. And also we saw the offspring of the giants there. 29. Amalek lives in the land of the Negeb, and the Hittite and the Jebusite and the Amorite live in the mountains, and the Canaanite lives by the sea and along the Jordan."

30. And Caleb quieted the people toward Moses and said, "*Let's go up,* and we'll take possession of it, because *we'll be able* to handle it."

31. And the people who went up with him said, "We won't be able to go up against the people, because they're stronger than we are. 33. And we saw the Nephilim there, sons of giants from the Nephilim, and we were like grasshoppers in our eyes, and so were we in their eyes."

14:1b. And the people wept that night, 4. and they said, each man to his brother, "Let's appoint a chief and go back to Egypt."

11. And YHWH said to Moses, "How long will this people reject me, and how long will they not trust in me, with all the signs that I've done among them? 12. I'll strike them with disease and dispossess them, and I'll make you into a bigger and more powerful nation than they are."

13. And Moses said to YHWH, "And Egypt will hear it, for you brought this people up from among them with your power, 14. and they'll say it to those who live in this land. They've heard that you, YHWH, are among this people; that you, YHWH, have appeared eye to eye; and your cloud stands over them; and you go in front of them in a column of cloud by day and in a column of 0fire by night. 15. And if you kill this people as one man, then the

nations that have heard about you will say it, saying, 16. 'Because YHWH wasn't able to bring this people to the land that He swore to them, He slaughtered them in the wilderness.' 17. And now, let my Lord's power be big, as you spoke, saying, 18. 'YHWH is slow to anger and abounding in kindness, bearing crime and offense, and not clearing: reckoning fathers' crime on children, on third generations and on fourth generations.' 19. Forgive this people's crime in proportion to the magnitude of your kindness and as you've borne this people from Egypt to here."

20. And YHWH said, "I've forgiven according to your word." 21. But indeed, as I live, and as YHWH's glory has filled all the earth, 22. I swear that all of these people, who have seen my glory and my signs that I did in Egypt and in the wilderness and who have tested me ten times now and haven't listened to my voice, 23. won't see the land that I swore to their fathers, and all those who rejected me won't see it. 24. And my servant Caleb, because a different spirit was with him, and he went after me completely, I'll bring him to the land where he went, and his seed will possess it. 25. And the Amalekite and the Canaanite live in the valley. Turn and travel tomorrow to the wilderness by the way of the Red Sea."

39. And Moses spoke these things to all of the children of Israel, and the people mourned very much. 40. And they got up in the morning and went up to the top of the mountain, saying, "Here we are, and we'll go up to the place that YHWH said, for we've sinned."

41. And Moses said, "Why are you violating YHWH's word? And it won't succeed. 42. Don't go up, or else you'll be stricken before your enemies, because YHWH isn't among you. 43. Because the Amalekite and the Canaanite are there before you, and you'll fall by the sword because of the fact that you've gone back from following YHWH, and YHWH won't be with you."

44. And they acted heedlessly, going up to the top of the mountain. And the ark of the covenant of YHWH and Moses did not draw away from the camp. 45. And the Amalekite and the Canaanite who lived in that mountain came down and struck as far as Hormah.

16:1b. And Dathan and Abiram, sons of Eliab, and On, son of Peleth, sons of Reuben, 2a. got up before Moses. 12. And Moses sent to call Dathan and Abiram, sons of Eliab, and they said, "We won't come up. 13. Is it a small thing that you brought us up from a land flowing with milk and honey to kill us in the wilderness, that you lord it over us as well? 14. Besides, you haven't brought us to a land flowing with milk and honey or given us possession of field or vineyard. Will you put out those people's eyes? We won't come up."

25. And Moses got up and went to Dathan and Abiram, and Israel's elders went after him. 27b. And Dathan and Abiram came out, standing at the entrance of their tents, and their wives and their children and their infants. 28. And Moses said, "By this you'll know that YHWH sent me to do all these things, because it's not from my own heart. 29. If these die like the death of every human, and the event of every human happens to them, then YHWH hasn't sent me. 30. But if YHWH will create something, and the ground will open its mouth and swallow them and all that they have, and they'll go down alive to Sheol, then you'll know that these people have rejected YHWH."

31. And it was as he was finishing speaking all these things, and the ground that was under them was split open, 32a. and the earth opened its mouth and swallowed them and their houses, 33. and they went down, they and all that they had, alive to Sheol. And the earth covered them over, and they perished from among the community. 34. And all Israel that was around them fled at the sound of them, for they said, "Or else the earth will swallow us."

20:14. And Moses sent messengers from Kadesh to the king of Edom. "Your brother, Israel, says this: 'You know all the hardship that has found us: 15. And our fathers went down to Egypt, and we lived in Egypt many days. And Egypt was bad to us and to our fathers, 16. and we cried to YHWH, and He heard our voice and sent an angel and brought us from Egypt. And here we are in Kadesh, a city at the edge of your border. 17. Let us pass through your land. We won't pass through a field or through a vineyard, and we won't drink the water of a well. We'll go by the king's road, we won't turn right or left, until we pass your border.'"

18. And Edom said to him, "You won't pass through me, or else I'll go out to you with the sword."

19. And the children of Israel said to him, "We'll go up by the highway, and if we drink any of your water, I and my cattle, then I'll pay its price. Only, it's nothing: let me pass through by foot."

20. And he said, "You won't pass." And Edom went out to him with a heavy mass of people and a strong hand, 21. and Edom refused to allow Israel to pass through his border. And Israel turned away from him.

21:1. And the Canaanite, the king of Arad, who lived in the Negeb, heard that Israel was coming by the way of Atharim, and he fought against Israel and took some of them prisoners. 2. And Israel made a vow to YHWH and said, "If you will deliver this people into my hand, then I shall completely destroy their cities." 3. And YHWH listened to Israel's voice and delivered the Canaanite, and they completely destroyed them and their cities. And the name of the place was called Hormah.

21. And Israel sent messengers to Sihon, king of the Amorites, saying, 22. "Let me pass through your land. We won't turn in at a

field and at a vineyard and won't drink well waters. We'll go by the king's road until we pass your border." 23. And Sihon did not allow Israel to pass within his border. And Sihon gathered all his people and went out to Israel at the wilderness and came to Jahaz and fought against Israel. 24. And Israel struck him with the sword and took possession of his land from Arnon to Jabbok, to the children of Ammon, because the border of the children of Ammon was strong. 25. And Israel took all of these cities, and Israel lived in all of the Amorite cities in Heshbon and all of its environs, 26. because Heshbon was the city of Sihon, the king of the Amorite, and he fought against the former king of Moab and took all of his land from his hand as far as Arnon. 27. On account of this they say proverbially:

Come to Heshbon.
Built and founded is Sihon's city.

28. For fire went out of Heshbon,
flame from Sihon's town;
it consumed Ar of Moab,
the lords of the high places of Arnon.

29. Woe unto you, Moab:
you've perished, Chemosh's people.
He's made his sons refugees
and his daughters captive
to the king of the Amorite, Sihon.

30. And their fiefdom perished,
Heshbon to Dibon,
and we devastated up to Nophah
which reaches to Medeba.

31. And Israel lived in the Amorite's land. 32. And Moses sent to spy on Jazer, and they captured its environs and dispossessed the Amorite who was there. 33. And they turned and went up the road of Bashan, and Og, the king of Bashan, went out to them, he and all of his people, to war at Edrei. 34. And YHWH said to Moses, "Don't fear him, because I've delivered him and all of his people and his land into your hand, and you'll do to him as you did to Sihon, king of the Amorite, who lived in Heshbon." 35. And they struck him and his sons and all of his people until he did not have any remnant left, and they took possession of his land.

25:1. And Israel lived in Shittim, and the people began to prostitute themselves to the daughters of Moab: 2. And they attracted the people to their gods' sacrifices, and the people ate and bowed to their gods, 3. and Israel was associated with Baal Peor. And YHWH's anger flared at Israel. 4. And YHWH said to Moses, "Take all of the leaders of the people and hang them in front of the sun, and YHWH's anger will go back from Israel."

5. And Moses said to Israel's judges, "Each of you, kill those of his people who are associated with Baal Peor."

Deuteronomy 34:5. And Moses, YHWH's servant, died there in the land of Moab by YHWH's mouth, 6. and He buried him in the valley in the land of Moab opposite Beth Peor. And no man knows his burial place to this day. 7. And Moses was a hundred twenty years old at his death. His eye was not dim, and his vigor had not fled.

Joshua 1:1. And it was after the death of Moses, YHWH's servant, and YHWH said to Joshua, son of Nun, Moses' attendant, saying, 6. "Be strong and bold, because you will enable this people to inherit the land that I swore to their fathers to give them. 9b. Don't shake and don't be dismayed, because YHWH, your God, is with you in every place that you go."

2:1. And Joshua, son of Nun, sent two spies from Shittim secretly, saying, "Go see the land, and specifically Jericho." And they went, and they came to a prostitute's house, and her name was Rahab. And they lay there.

2. And it was said to the king of Jericho, saying, "Here, men have come here tonight from the children of Israel to probe the land."

3. And the king of Jericho sent to Rahab, saying, "Bring out the men who have come to you, who came to your house, because they came to probe all of the land."

4. And the woman took the two men and concealed them, and she said, "Yes, the men came to me, and I didn't know where they were from. 5. And the gate was about to close at dark, and the men went out. I don't know where the men went. Pursue them quickly so you'll overtake them." 6. And she had taken them up to the roof and hidden them among the stalks of flax that she had stacked on the roof. 7. And the people pursued them toward the Jordan at the fords. And they closed the gate after the pursuers went out after them. 8. And they had not yet lain down, and she came up to them on the roof 9. and said to the men, "I know that YHWH has given you the land, because terror of you has fallen on us and because everyone living in the land is dissolving before you, 10. because we've heard that YHWH dried the waters of the Red Sea before you when you went out of Egypt and what you did to the two Amorite kings who were across the Jordan, to Sihon and Og, that you completely destroyed them. 11. And we heard, and our heart melted, and no more spirit rose in any man before you, because YHWH, your God, He is God, in the skies

above and on the earth below. 12. And now, swear to me by YHWH, because I've practiced kindness with you, that you, too, will practice kindness with my father's house and give me a sign of trust 13. and will keep my father and my mother and my brothers and my sisters and all that they have alive, and will rescue our lives from death."

14. And the men said to her, "Our soul to die in place of yours—if you don't tell this business of ours; and it will be, when YHWH gives us this land, that we'll practice kindness and faithfulness with you."

15. And she let them down by a rope through the window because her house was in the wall of the rampart, and she lived in the rampart. 16. And she said to them, "To the mountain! Go! In case the pursuers run into you. And hide there for three days, until the pursuers come back, and then go your way."

17. And the men said to her, "We're free from this oath of yours that you've had us swear: 18. Here we're coming into the land. Tie this cord of scarlet rope in the window by which you let us down, and gather your father and your mother and your brothers and all of your father's house to you at the house. 19. And anyone who'll go out from the doors of your house to the outside, his blood will be on his hands, and we'll be free; and anyone who'll be with you in the house, his blood is on our head if a hand will be laid on him. 20. And if you tell this business of ours, then we'll be free from your oath that you've had us swear."

21. And she said, "According to your words, so be it." And she sent them away, and they went. And she tied the scarlet cord in the window.

22. And they went, and they came to the mountain and stayed there three days, until the pursuers went back, and the pursuers sought all along the way and did not find them. 23. And the two men came back: and they came down from the mountain and

crossed and came to Joshua, son of Nun, and told him all the things that had happened to them. 24. And they said to Joshua, "YHWH has delivered all the land into our hand; and, also, everyone living in the land is dissolving before us."

3:1. And Joshua got up early in the morning, and they departed from Shittim and came to the Jordan, he and all the children of Israel, and they spent the night there before they crossed. 2. And it was at the end of three days, and the officers passed through the camp 3. and commanded the people, saying, "As you see the ark of the covenant of YHWH, your God, and the Levite priests carrying it, then you shall depart from your place and go behind it. 4. Only there shall be a distance between you and it, about two thousand cubits in measure. Don't come close to it, so that you'll know the way in which you'll go, because you've never passed this way in previous days."

5. And Joshua said to the people, "Consecrate yourselves, because tomorrow YHWH will do wonders among you." 6. And Joshua said to the priests, saying, "Carry the ark of the covenant and pass before the people." And they carried the ark of the covenant and passed before the people.

7. And YHWH said to Joshua, "This day I'll begin making you great in the eyes of all of Israel so that they'll know that as I was with Moses I'll be with you. 8. And you shall command the priests carrying the ark of the covenant, saying, 'When you come to the edge of the Jordan's waters, you shall stand in the Jordan.'"

9. And Joshua said to the children of Israel, "Come over here and hear the words of YHWH, your God." 10. And Joshua said, "By this you'll know that a living God is among you and will *dispossess* the Canaanite and the Hittite and the Hivite and the Perizzite and

the Girgashite and the Amorite and the Jebusite on your account: 11. Here is the ark of the covenant of the Lord of all the earth passing in front of you in the Jordan. 12. And now, take twelve men from the tribes of Israel, one man for each tribe, 13. and it will be, when the feet of the priests carrying the ark of YHWH, Lord of all the earth, rest in the Jordan's waters, the Jordan's waters will be cut off, the waters that come down from above, and will stand as one heap." 14. And it was when the people traveled from their tents to cross the Jordan—and the priests carrying the ark of the covenant in front of the people—15. and when the ones carrying the ark came to the Jordan, and the legs of the priests carrying the ark were dipped in the edge of the water (as the Jordan flows over all its banks all the days of harvest) 16. that the waters that were coming down from above stood; they rose as one heap, very far from Adam, the city that is by Zaretan. And those that were going down toward the sea of Arabah, the Salt Sea, had finished; they were cut off. And the people crossed opposite Jericho. 17. And the priests who were carrying the ark of the covenant of YHWH stood on dry ground in the Jordan firmly while all of Israel had been crossing on dry ground, until all of the nation finished crossing the Jordan. 4:14. In that day YHWH made Joshua great in the eyes of all of Israel, and they feared him as they had feared Moses all the days of his life.

5:1. And it was when all the kings of the Amorite who were on the west side the Jordan and all the kings of the Canaanite who were by the sea heard that YHWH had dried the Jordan's waters in front of the children of Israel until they had crossed: and their heart melted, and they had no more spirit before the children of Israel.

2. At that time YHWH said to Joshua, "Make flint knives and go back, circumcise the children of Israel a second time." 3. And Joshua made flint knives and circumcised the children of Israel at the Hill of the Foreskins. 4. And this is the reason that Joshua circumcised: All

of the people who had gone out of Egypt, the males, all the men of war, died on the way in the wilderness when they had gone out of Egypt, 5. because all of the people who had gone out were circumcised, and all of the people who were born on the way in the wilderness when they had gone out of Egypt had not been circumcised. 6. For the children of Israel went in the wilderness for forty years, until the whole nation came to an end, the men of war who had gone out of Egypt, who had not listened to YHWH's voice, to whom YHWH had sworn that He would not have them see the land that YHWH had sworn to their fathers to give us, a land flowing with milk and honey. 7. And He established their children in their place. Joshua circumcised them, because they were uncircumcised, because they did not circumcise them on the way. 8. And it was when all the nation had ended being circumcised: and they stayed in their places in the camp until they revived. 9. And YHWH said to Joshua, "Today I've rolled the disgrace of Egypt off of you." And he called that place's name Gilgal to this day.

13. And it was when Joshua was at Jericho: and he raised his eyes and saw, and here was a man standing opposite him, and his sword was drawn in his hand. And Joshua went to him and said to him, "Are you for us or for our enemies?"

14. And he said, "No. For I'm commander of YHWH's army. Now I've come."

And Joshua fell on his face to the ground and bowed and said to him, "What does my lord speak to his servant?"

15. And the commander of YHWH's army said to Joshua, "Take off your shoe from your foot, because the place on which you're standing: it's holy." And Joshua did so.

6:1. And Jericho was utterly closed because of the children of Israel: no one going out and no one coming in. 2. And YHWH said to Joshua, "See, I've delivered Jericho and its king, powerful warriors, into your hand. 3. And you shall go around the city, all the men of war, circling the city one time. You shall do that for six days. 4. And seven priests will carry seven rams' horns that are for blowing before the ark. And on the seventh day you shall go around the city seven times, and the priests will blast on the rams' horns. 5. And it will be at the horn's prolonged blowing, when you hear the sound of the ram's horn, all the people shall make a big shout: and the city's rampart will collapse in place. And the people shall go up, each one straight before him."

6. And Joshua, son of Nun, called the priests and said to them, "Carry the ark of the covenant, and let seven priests carry seven rams' horns, blowing before YHWH's ark." 7. And he said to the people, "Pass on and go around the city, and let an armed group pass in front of YHWH's ark." 8. And it was as Joshua said it to the people, and seven priests carrying seven rams' horns for blowing before YHWH passed and blasted on the rams' horns, and the ark of the covenant of YHWH was going after them. 9. And the armed group was going in front of the priests who were blasting on the rams' horns, and the rear guard was going behind the ark—with the continual blasting on the rams' horns. 10. And Joshua commanded the people, saying, "You shall not shout, and you shall not let your voices be heard, and not a word will go out from your mouth until the day I say to you, 'Shout!' Then you'll shout." 11. And YHWH's ark went around the city, circling one time, and they came to the camp and spent the night in the camp.

12. And Joshua got up early in the morning, and the priests carried YHWH's ark, 13. and the seven priests carrying seven rams' horns for blowing before YHWH's ark were going, and they

blasted on the rams' horns continually, and the armed group was going in front of them and the rear guard going behind YHWH's ark with the continual blasting on the rams' horns. 14. And they went around the city on the second day one time and came back to the camp. They did that for six days.

15. And it was on the seventh day, and they got up early, at the rising of the morning light, and went around the city in this manner seven times; only on that day they went around the city seven times. 16. And it was on the seventh time: the priests blasted the rams' horns, and Joshua said to the people, "Shout! because YHWH has given you the city. 17. And the city shall be a complete destruction, it and everything in it, to YHWH. Only Rahab the prostitute shall live, she and everyone who is with her in the house, because she hid the messengers whom we sent. 18. Only beware of the rule of complete destruction, in case you become subject to it if you take anything from what is to be completely destroyed and you subject the camp of Israel to complete destruction and cause it anguish. 19. And all silver and gold and implements of bronze and iron are consecrated to YHWH; they shall come to YHWH's treasury."

20. And the people shouted. And they blasted on the rams' horns. And it was when the people heard the sound of the ram's horn: and the people made a big shout. And the rampart collapsed in place. And the people went up to the city, each one straight before him, and they captured the city, 21. and they completely destroyed everything that was in the city, from man to woman, from young to old, and to ox and sheep and ass, by the edge of sword. 22. And Joshua said to the two men who had spied out the land, "Come to the prostitute woman's house and bring out the woman and all she has from there as you swore to her." 23. And the young men who had spied out came and brought out Rahab

and her father and her mother and her brothers and all she had. And they brought out all her relatives and set them down outside the camp of Israel. 24. And they burned the city and everything in it in fire; only the silver and the gold and the implements of bronze and iron they put in the treasury of the house of YHWH. 25. And Joshua kept Rahab the prostitute and her father's house and all she had alive, and she has lived among Israel to this day because she hid the messengers whom Joshua sent to spy out Jericho. 26. And Joshua swore at that time, saying, "Let the man who arises and builds this city, Jericho, be cursed before YHWH. He'll found it with his firstborn and set up its doors with his youngest."

27. And YHWH was with Joshua, and his fame was in all the land.

7:1. And the children of Israel had made a breach in the rule of complete destruction. And Achan, son of Carmi, son of Zabdi, son of Zerah, of the tribe of Judah, had taken from what was to be completely destroyed. And YHWH's anger flared at the children of Israel.

2. And Joshua sent people from Jericho to Ai, which is by Beth-aven, east of Beth-El, and said to them, saying, "Go up and spy out the land." And the people went up and spied on Ai 3. and came back to Joshua and said to him, "Let not all the people go up. Let around two thousand men, or around three thousand men, go up and strike Ai. Don't burden all the people to go there, because they're few." 4. And around three thousand men from the people went up there, and they fled before the people of Ai. 5. And the people of Ai struck around thirty-six men of them and pursued them before the gate as far as Shebarim and struck them in the descent.

And the people's heart melted and became water. 6. And Joshua ripped his clothes and fell on his face to the ground before YHWH's ark until the evening, he and Israel's elders, and they put dust on their heads. 7. And Joshua said, "Oh, Lord YHWH, why did you *bring* this people across the Jordan to deliver us into the Amorite's hand to destroy us? And if only we'd been content and lived beyond the Jordan. 8. Please, my Lord, what shall I say after Israel has turned its back before its enemies? 9. And the Canaanite and everyone who lives in the land will hear and go around us and cut off our name from the earth. And what will you do for your great name?"

10. And YHWH said to Joshua, "Get up. What is this, that you're falling on your face? 11. Israel has sinned, and violated my covenant that I commanded them as well, and taken from what was to be completely destroyed as well, and stolen as well, and covered up as well, and put it among their possessions as well. 12. And the children of Israel weren't able to stand up before their enemies, they turned their backs before their enemies, because they had *become* what was to be completely destroyed. I won't continue to be with you if you don't eliminate what was to be completely destroyed from among you. 13. Get up. Consecrate the people. And say, "Consecrate yourselves, because YHWH, Israel's God, has said this: 'Some of what was to be completely destroyed is among you, Israel. You won't be able to stand up before your enemies until you remove what was to be completely destroyed from among you.' 14. And you will come forward in the morning by your tribes, and it will be that the tribe that YHWH will take by lot will come forward by families, and the family that YHWH will take by lot will come forward by households, and the household that YHWH will take by lot will come forward by individuals. 15. And it will be that the one who is taken by lot with what was to be

completely destroyed will be burned in fire, he and everything he has, because he violated YHWH's covenant, and because he did a foolhardy thing among Israel."

16. And Joshua got up early in the morning and brought Israel forward by its tribes: and the tribe of Judah was taken by lot. 17. And he brought Judah's families forward: and the Zerahite family was taken by lot. And he brought the Zerahite family forward by men: and Zabdi was taken by lot. 18. And he brought his house forward by men, and Achan, son of Carmi, son of Zabdi, son of Zerah, of the tribe of Judah was taken by lot.

19. And Joshua said to Achan, "My son, accord glory to YHWH, Israel's God, and give him acknowledgment, and tell me what you did. Don't hide it from me."

20. And Achan answered Joshua and said, "Truly, I've sinned against YHWH, Israel's God, and I've done like this and like this: 21. and I saw a robe from Shinar in the booty, a good one, and two hundred shekels of silver and one bar of gold weighing fifty shekels, and I coveted them, and I took them. And they're here hidden in the ground inside my tent, and the silver is under it."

22. And Joshua sent messengers, and they ran to the tent, and here it was hidden in his tent, and the silver was under it. 23. And they took them from the tent and brought them to Joshua and to all the children of Israel, and they set them down before YHWH. 24. And Joshua took Achan, son of Zerah, and the silver and the robe and the bar of gold and his sons and his daughters and his ox and his ass and his flock and his tent and everything he had; and all Israel was with him, and they brought them up to the valley of Achor. 25. And Joshua said, "Why did you cause us anguish? YHWH will cause *you* anguish on this day." And all Israel battered him with stones and burned them with fire and stoned them with stones, 26. and they set up a big pile of stones over him to this day. And

YHWH turned back from His flaring anger. On account of this the name of that place was called the valley of Achor to this day.

8:1. And YHWH said to Joshua, "Don't be afraid and don't be dismayed. Take all the people of war with you and get up, go to Ai. See, I've delivered the king of Ai and his people and his city and his land into your hand, 2. and you'll do to Ai and its king as you did to Jericho and its king; only you'll despoil its booty and its cattle for yourselves. Set an ambush for the city from behind it."

3. And Joshua and all the people of war rose to go up to Ai, and Joshua chose thirty thousand men, powerful warriors, and sent them by night. 4. And he commanded them, saying, "See, you're ambushing the city from behind the city. Don't go very far from the city. And, all of you, be ready. 5. And I and all the people who are with me will go near to the city, and it will be, when they come out toward us as on the first time, that we'll flee before them. 6. And they'll come out after us until we've drawn them away from the city, because they'll say, 'They're fleeing before us as on the first time,' and we'll flee before them. 7. And you'll get up from the ambush and take possession of the city, and YHWH, your God, will deliver it into your hand. 8. And it will be, when you seize the city: you'll set the city on fire. According to YHWH's word you'll do it. See, I've commanded you." 9. And YHWH sent them, and they went to the ambush and stayed between Beth-El and Ai, to the west of Ai. And Joshua spent that night among the people.

10. And Joshua got up early in the morning and mustered the people, and he and Israel's elders went up to Ai in front of the people. 11. And all the people of war who were with him went up and approached and came opposite the city and camped to the north of Ai, and a valley was between them and Ai. 12. And he took around five thousand men and set them in ambush between Beth-El and Ai, to the west of Ai. 13. And they set the people—all of the camp that

was to the north of the city and their ambush group to the west of the city. And Joshua went through the valley that night. 14. And it was when the king of Ai saw it: and the people of the city hurried and got up early and went out toward Israel for war, he and all of his people at the set time facing the Arabah. And he did not know that there was an ambush for him behind the city. 15. And Joshua and all of Israel acted beaten before them and fled by the way of the wilderness. 16. And all the people who were in the city cried out to pursue them, and they pursued Joshua and were drawn from the city. 17. And not a man was left in Ai and Beth-El who had not gone out after Israel, and they left the city open and pursued Israel.

18. And YHWH said to Joshua, "Reach out with the spear that's in your hand toward Ai, because I'll deliver it into your hand." And Joshua reached out with the spear that was in his hand toward the city. 19. And the ambush rose quickly from its place, and they ran as his hand reached out, and they came to the city and captured it. And they hurried and set the city on fire. 20. And the people of Ai looked behind them and saw; and, here, the smoke of the city went up to the skies. And they did not have outlets to flee this way or that way. And the people who had fled to the wilderness were turned into the pursuer. 21. And Joshua and all Israel saw that the ambush had captured the city and that the city's smoke went up, and they went back and struck the people of Ai, 22. and these came out from the city toward them so that they were in between Israel—these from this side and these from that side—and they struck them until they did not have any remnant or escapee left. 23. And they took the king of Ai alive and brought him forward to Joshua. 24. And it was as Israel ended killing everyone who lived in Ai, in the field, in the wilderness where they had pursued them, and they had all fallen by the sword until they were finished, and all Israel came back to Ai and struck it by the sword. 25. And

all those who fell in that day, from man to woman, were twelve thousand, all the people of Ai. 26 And Joshua did not pull back his hand that he had reached out with the spear until he had completely destroyed everyone who lived in Ai. 27. Only Israel despoiled for themselves the cattle and that city's booty, according to YHWH's word which He had commanded Joshua. 28. And Joshua burned Ai and made it a mound forever, a desolation to this day. 29. And he hanged the king of Ai on a tree until the evening, and when the sun set Joshua gave a command, and they took his corpse down from the tree and threw it down at the entrance of the city's gate and set up a big pile of stones over him to this day.

9:1. And it was when all the kings who were across the Jordan heard—in the hills and in the lowland and in all the coast of the big sea up to Lebanon, the Hittite and the Amorite, the Canaanite, the Perizzite, the Hivite, and the Jebusite—2. and they gathered together, of one voice, to fight against Joshua and Israel.

3. And those who lived in Gibeon heard what Joshua had done to Jericho and Ai, 4. and they, too, made a ruse: and they went and took provisions and took worn sacks for their asses and worn and torn and mended wineskins 5. and worn and spotted shoes on their feet and worn clothes on them, and all the bread in their provisions was dry, speckled. 6. And they went to Joshua, to the camp at Gilgal, and said to him and to the men of Israel, "We've come from a distant land, and now make a covenant with us."

7. And the men of Israel said to the Hivite, "Maybe you live in my region, and how shall I make a covenant with you?"

8. And they said to Joshua, "We're your servants."

And Joshua said to them, "Who are you, and from where do you come?"

9. And they said to him, "Your servants have come from a very distant land to the name of YHWH, your God, because we heard about Him and all that He did in Egypt 10. and all that He did to the two Amorite kings who were beyond the Jordan, to Sihon, king of Heshbon, and to Og, king of Bashan, who was in Ashtarot. 11. And our elders and everyone living in our land said to us, saying, 'Take food in your hand for the road and go to them, and you shall say to them, "We're your servants, and now make a covenant with us."' 12. This is our bread. We took it hot as provisions from our houses on the day we went out to come to you, and now here it is: dry, and it has become speckled. 13. And these are the wineskins that we filled when they were new, and here they've broken up. And these are our clothes and our shoes; they've worn out from the very long road."

14. And the people took some of their food. And they did not ask from YHWH's mouth. 15a. And Joshua made peace with them and made a covenant with them to keep them alive.

16. And it was after they had made a covenant with them: and they heard that they were close to them and they lived among them. 22. And Joshua called to them and spoke to them, saying, "Why did you deceive us, saying, 'We're very distant from you,' when you live among us?! 23. And now you're cursed, and none of you will be cut loose from being a servant—and woodcutters and water-drawers for my God's house."

24. And they answered Joshua and said, "Because it was *told* to your servants what YHWH, your God, commanded Moses, His servant: to give you all of the land and to destroy everyone who lives in the land before you. And we were very afraid for our lives before you, and we did this thing. 25. And now, here we are in your hand. Do whatever is good and right in your eyes to do to us."

26a. And he did that to them.

10:1. And it was when Adoni-zedek, king of Jerusalem, heard that Joshua had captured Ai and completely destroyed it—he had done to Ai and its king just as he had done to Jericho and its king—and that those who lived in Gibeon had made peace with Israel and were among them, 2. and they were very afraid because Gibeon was a big city, like one of the royal cities, and because it was bigger than Ai, and all its people were powerful warriors. 3. And Adoni-zedek, king of Jerusalem, sent to Hoham, king of Hebron, and to Piram, king of Jarmuth, and to Japhia, king of Lachish, and to Debir, king of Eglon, saying, 4. "Come up to me and help me, and we'll strike Gibeon because it made peace with Joshua and with the children of Israel."

5. And they gathered and went up, the five kings of the Amorites: the king of Jerusalem, the king of Hebron, the king of Jarmuth, the king of Lachish, the king of Eglon, they and all their camps, and encamped at Gibeon and fought against it.

6. And the people of Gibeon sent to Joshua, to the camp at Gilgal, saying, "Don't hold off your hand from your servants. Come up to us quickly and save us and help us because all the kings of the Amorite who live in the mountains have gathered against us." 7. And Joshua went up from Gilgal, he and all the people of war with him and all the powerful warriors.

8. And YHWH said to Joshua, "Don't be afraid of them, because I've put them in your hand. Not a man of them will stand up against you."

9. And Joshua came to them suddenly. He'd gone up from Gilgal all night. 10. And YHWH threw them into tumult before Israel. And he struck them a big blow at Gibeon and pursued them by way of the ascent of Beth-Horon and struck them as far as Azekah and as far as Makkedah. 11. And it was when they were fleeing

from Israel: they were in the descent of Beth-Horon, and YHWH threw big stones down on them from the skies as far as Azekah, and they died. There were more who died in the hailstones than those whom the children of Israel killed by the sword.

12. Then Joshua spoke to YHWH, in the day that YHWH delivered the Amorite before the children of Israel, and before Israel's eyes he said:

Sun stand still in Gibeon
and moon in the valley of Ayyalon!
13. And the sun stood still and the moon stood
until a nation requited its enemies.

Is it not written in the book of Yashar? And the sun stood in the middle of the skies and did not move quickly to set for a whole day.

14. And there was nothing like that day, before it and after it, regarding YHWH's listening to a man's voice, because YHWH fought for Israel.

15. And Joshua came back, and all Israel with him, to the camp at Gilgal. 16. And these five kings fled and were hidden in a cave in Makkedah. 17. And it was told to Joshua, saying, "The five kings have been found hidden in a cave in Makkedah."

18. And Joshua said, "Roll big stones to the mouth of the cave and post people over it to watch them. 19. And you, don't stand. Pursue your enemies and tail them. Don't allow them to come to their cities, because YHWH, your God, has delivered them into your hand."

20. And it was when Joshua and the children of Israel finished striking them a very big blow until they were finished: and the remnants broke away from them and came to the fortified cities. 21. And all the people came back to the camp, to Joshua, at Makkedah in peace; no one moved his tongue against the children

of Israel. 22. And Joshua said, "Open the mouth of the cave and bring these five kings out of the cave to me." 23. And they did that: and they brought these five kings out of the cave to him: the king of Jerusalem, the king of Hebron, the king of Jarmuth, the king of Lachish, the king of Eglon. 24. And it was when they brought these kings out to Joshua: and Joshua called to all the men of Israel and said to the officers of the men of war who had gone with him, "Come forward. Put your feet on these kings' necks." And they came forward and put their feet on their necks. 25. And Joshua said to them, "Don't be afraid and don't be dismayed. Be strong and bold because YHWH will do likewise to all your enemies against whom you're fighting." 26. And Joshua struck them after that and killed them and hanged them on five trees. And they were hanging on the trees until evening. 27. And it was at the time of the setting of the sun that Joshua gave the command, and they brought them down from the trees and threw them into the cave in which they had been hidden, and they put big stones over the mouth of the cave to this very day.

28. And Joshua captured Makkedah in that day and struck it and its king with the edge of the sword. He completely destroyed them and every soul that was in it. He did not leave a remnant. And he did to the king of Makkedah as he had done to the king of Jericho. 29. And Joshua passed, and all Israel with him, from Makkedah to Libnah and fought with Libnah. 30. And YHWH delivered it, too, and its king into Israel's hand, and he struck it and every soul that was in it with the edge of the sword. He did not leave a remnant in it. And he did to its king as he had done to the king of Jericho. 31. And Joshua passed, and all Israel with him, from Libnah to Lachish and camped against it and fought against it. 32. And YHWH delivered Lachish into Israel's hand, and he captured it on the second day and struck it and every soul that was in it with the edge of the sword,

like everything that he had done to Libnah. 33. Then Horam, king of Gezer, went up to help Lachish, and Joshua struck him and his people until he did not have any remnant left. 34. And Joshua passed, and all Israel with him, from Lachish to Eglon and camped against it and fought against it. 35. And they captured it in that day and struck it with the edge of the sword, and he completely destroyed every soul that was in it, like everything that he had done to Lachish. 36. And Joshua went up, and all Israel with him, from Eglon to Hebron, and they fought against it 37. and captured it and struck it and its king and all its cities and every soul that was in it with the edge of the sword. He did not leave a remnant, like everything he had done to Eglon, and he completely destroyed it and every soul that was in it. 38. And Joshua came back, and all Israel with him, to Debir, and he fought against it 39. and captured it and its king and all its cities, and they struck them with the edge of the sword and completely destroyed every soul that was in it. He did not leave a remnant. As he had done to Hebron, so he did to Debir and its king, and as he had done to Libnah and its king. 40. And Joshua struck all the land: the mountain and the Negeb and the lowland and the slopes, and all their kings. He did not leave a remnant, and he completely destroyed everything that breathed, as YHWH, Israel's God, had commanded. 41. And Joshua struck them from Kadesh Barnea to Gaza, and all the land of Goshen and to Gibeon. 42. And Joshua captured all these kings and their land at one time because YHWH, Israel's God, was fighting for Israel. 43. And Joshua went back, and all Israel with him, to the camp at Gilgal.

11:1. And it was when Jabin, king of Hazor, heard: and he sent to Jobab, king of Madon, and to the king of Shimron and to the king of Achshaph 2. and to the kings who were in the north, in the mountains and in the Arabah south of Chinerot and in the lowlands and in the areas of Dor in the west, 3. the Canaanite in the east and in the west, and the Amorite and the Hittite and the Perizzite and the Jebusite in the mountains and the Hivite below Hermon in the land of Mizpah. 4. And they went out, they and all their camps with them, a great number of people, like the sand that is on the seashore for its great number, and very many horses and chariots. 5. And all these kings met and came and encamped together at the waters of Merom to fight against Israel.

6. And YHWH said to Joshua, "Don't be afraid of them, because at this time tomorrow I'm delivering all of them slain before Israel. You shall hamstring their horses and burn their chariots with fire."

7. And Joshua and all the people of war with him came upon them at the waters of Merom suddenly and fell upon them. 8. And YHWH delivered them into Israel's hand, and they struck them and pursued them to great Sidon and to Misrephot-maim and to the valley of Mizpeh in the east. And they struck them until he did not leave them a remnant. 9. And Joshua did to them as YHWH had said to him: he hamstrung their horses and burned their chariots with fire.

10. And Joshua went back at that time and captured Hazor and struck its king with the sword because Hazor formerly had been the head of all these kingdoms. 11. And they struck every soul in it with the edge of the sword, completely destroying; none who breathed was left, and they burned Hazor with fire. 12. And Joshua captured all these kings' cities and all their kings, and he struck them with the edge of the sword. He completely destroyed them, as Moses, YHWH's servant, had commanded. 13. Only, all the cities that are standing on their mounds: Israel did not burn them, but Joshua burned Hazor alone. 14. And the children of Israel despoiled for themselves all the booty of these cities and the cattle; they struck only all the humans with the edge of the sword until they had destroyed them. They did not leave any who breathed.

15. As YHWH had commanded Moses, His servant, so Moses commanded Joshua, and so Joshua did. He did not omit a thing out of all that YHWH had commanded Moses. 16. And Joshua took all this land: the mountains and all of the Negeb and all the land of Goshen and the lowland and the Arabah and Israel's mountains and its lowland. 17. He captured from Mount Halak, which goes up to Seir, to Baal-Gad in the valley of Lebanon below Mount Hermon and all their kings, and he struck them and killed them. 18. Joshua made war against all these kings many days. 19. There was no city that made peace with Israel except the Hivite, those living in Gibeon. They took everything in war. 20. For it was from YHWH, to strengthen their heart toward war with Israel in order to destroy them completely so that they would not have any favor, but so that he would destroy them as YHWH had commanded Moses.

21. And Joshua came at that time and cut off the giants from the mountains, from Hebron, from Debir, from Anab, and from all the mountains of Judah and from all the mountains of Israel. Joshua completely destroyed them with their cities. 22. No giants were left in the land of the children of Israel. Only in Gaza, in Gath, and in Ashdod did they remain.

23. And Joshua took all of the land according to all that YHWH had spoken to Moses, and Joshua gave it to Israel as a legacy according to their tribal divisions. And the land had respite from war.

13:1. And Joshua was old, well along in days, and YHWH said to him, "You've grown old, you're well along in days, and very much of the land remains to be possessed. 2. This is the remaining land: all the regions of the Philistines and all those of the Geshurite, 3. from Shihor, which is facing Egypt and as far as the border of Ekron to the north, reckoned to the Canaanite: five lords of the Philistines, the Gazaite and the Ashdodite, the Ashkelonite, the Gittite, and the Ekronite—and the Avites 4. in the south, all the land of the Canaanite, and Mearah, which the Sidonians have, to Aphek, to the Amorite border, 5. and the land of the Gebalite and all Lebanon toward the sunrise, from Baal-Gad below Mount Hermon to the entrance of Hamath. 6. All who live in the mountains from Lebanon to Misrephot-maim, all the Sidonians: I shall dispossess them before the children of Israel. Only allot it to Israel as a legacy as I have commanded you. 7. And now, divide this land for a legacy to the nine tribes and half the tribe of Manasseh." 8. The Reubenite and the Gadite took their legacy with it, which Moses gave them beyond the Jordan to the east, as Moses, YHWH's servant, had given them: 9. from Aroer, which is on the rim of the wadi Arnon, and the city that is inside the wadi, and all the plain of Medeba to Dibon, 10. and all the cities of Sihon, king of the Amorite, who ruled in Heshbon, to the border of the children of Ammon, 11. and Gilead and the Geshurite and Maachathite border and all of Mount Hermon and all of Bashan to Salcah, 12. all the kingdom of Og in Bashan, who ruled in Ashtarot and in Edrei—he was left from what remained of the Rephaim. Moses struck them and dispossessed them. 13. And the children of Israel did not dispossess the Geshurite and the Maachathite, and Geshur and Maachat have lived among Israel to this day.

Judges 8:30. And Jerubbaal had seventy sons issuing from his thigh because he had many wives. 31. And his concubine who was in Shechem also gave birth to a son for him, and he made his name

Abimelek. 32. And Jerubbaal died at a good old age and was buried in his father Joash's tomb in Ophrah of the Abiezrites. 9:1. And Abimelek, son of Jerubbaal, went to Shechem, to his mother's brothers, and spoke to them and to all the family of the house of his mother's father, saying, 2. "Speak in the ears of all of Shechem's citizens: 'What is good for you? Will seventy men, all of Jerubbaal's sons, rule you? Or will one man rule you?' And remember that I am your bone and your flesh."

3. And his mother's brothers spoke all of these things about him in the ears of all of Shechem's citizens, and their heart inclined to Abimelek because, they said, "He's our brother." 4. And they gave him seventy weights of silver from the house of Baal-berit, and Abimelek hired worthless and reckless people with them, and they followed him. 5. And he came to his father's house at Ophrah and killed his brothers, Jerubbaal's sons, seventy men, on one stone. And Jotham, Jerubbaal's youngest son, was left because he had been hidden.

6. And all of Shechem's citizens and all of the house of Millo gathered and went and made Abimelek king at the oak of the pillar that was in Shechem.

7. And they told Jotham, and he went and stood at the top of Mount Gerizim and raised his voice and called, and he said to them, "Listen to me, citizens of Shechem, and God will listen to you: 8. The trees went to anoint a king over them! And they said to the olive, 'Rule over us.'

9. "And the olive said to them, 'Should I stop producing my oil, by which they honor God and people through me, and go roaming over the trees?!'

10. "And the trees said to the fig, 'You, come. Rule over us.'

11. "And the fig said to them, 'Should I stop producing my sweetness and my good fruit and go roaming over the trees?!'

12. "And the trees said to the vine, 'You, come. Rule over us.'

13. "And the vine said to them, 'Should I stop producing my wine, which makes God and people happy, and go roaming over the trees?!'

14. "And all of the trees said to the bramble, 'You, come. Rule over us.'

15. "And the bramble said to the trees, 'If you're really anointing me as king over you, come, trust in my shade; and if not, let fire come out of the bramble and consume the cedars of Lebanon.'

16. "And now, if you've acted with truth and integrity in making Abimelek king, and if you've done right by Jerubbaal and his house, and if you've done to him what his hands have earned—18. because you've risen against my father's house today and killed his sons, seventy men, on one stone and made Abimelek, a son of his maid, king over Shechem's citizens because he's your brother—19. and if you've acted with truth and integrity toward Jerubbaal and his house this day, then be happy with Abimelek and let him be happy with you as well. 20. And if not, let fire come out of Abimelek and consume Shechem's citizens and the house of Millo, and let fire come out of Shechem's citizens and from the house of Millo and consume Abimelek." 21. And Jotham fled and ran away and went to Beer and lived there because of Abimelek, his brother.

22. And Abimelek had command of Israel three years. 23. And God sent a bad spirit between Abimelek and Shechem's citizens, and Shechem's citizens betrayed Abimelek, 24. so that the violence to the seventy sons of Jerubbaal would come upon, and their blood be placed upon, Abimelek, their brother who killed them, and on Shechem's citizens, who strengthened his hands to kill his brothers. 25. And Shechem's citizens set ambushers against him on the mountaintops, and they robbed everyone who passed by them on the road. And it was told to Abimelek.

26. And Gaal, son of Ebed, and his brothers came and passed through Shechem, and Shechem's citizens trusted him. 27. And they went out to the field and cut off and trod the grapes from their vineyards and made festivities, and they came to the house of their god and ate and drank and cursed Abimelek. 28. And Gaal, son of Ebed, said, "Who is Abimelek, and who is Shechem, that we should serve him? Didn't Jerubbaal's son and his officer Zebul serve the people of Hamor, Shechem's father? And why should *we* serve him? 29. And who would make it so that this people would be in my hand, and I would turn Abimelek out?!" And he said to Abimelek, "Build up your army and come out!"

30. And Zebul, the city's chief officer, heard the words of Gaal, son of Ebed, and his anger flared. 31. And he sent messengers to Abimelek at Tormah saying, "Here are Gaal, son of Ebed, and his brothers coming to Shechem, and, here, they're besieging the city against you. 32. And now, come up by night, you and the people with you, and set an ambush in the field. 33. And it will be that you'll get up early, in the morning at sunrise, and charge the city; and, here, he and the people with him will be coming out against you, and you'll do to him whatever your hand can achieve."

34. And Abimelek and all the people with him came up by night and set four ambush groups against Shechem. 35. And Gaal, son of Ebed, came out and stood at the entrance of the city gate. And Abimelek and the people with him came up from the ambush.

36. And Gaal saw the people and said to Zebul, "Here's a people coming down from the mountaintops."

And Zebul said to him, "You're seeing the mountains' shadow looking like people."

37. And Gaal went on to speak again and said, "Here are people coming down from the center of the land, and one group is coming from the direction of Elon Meonenim."

38. And Zebul said to him, "Where, oh where, is your mouth, with which you said, 'Who is Abimelek that we should serve him?' Isn't this the people whom you rejected? Go out now and fight against him." 39. And Gaal went out in front of the people of Shechem and fought against Abimelek. 40. And Abimelek pursued him, and he fled from him, and many fell, dead bodies, up to the city gate. 41. And Abimelek lived in Arumah, and Zebul drove out Gaal and his brothers from living in Shechem.

42. And it was the following day, and the people went out to the field. And they told Abimelek, 43. and he took the people and divided them into three groups and set an ambush in the field. And he saw, and here was the people coming out of the city, and he came up against them and struck them. 44. And Abimelek and the groups with him charged and stood at the entrance of the city gate, and the two groups charged everyone who was in the field and struck them. 45. And Abimelek fought against the city all that day and captured the city and killed the people in it and demolished the city and sowed it with salt.

46. And all the citizens of Shechem Tower heard about it and came to the stronghold of the house of El Berit. 47. And it was told to Abimelek that all the citizens of Shechem Tower had gathered together, 48. and Abimelek went up to Mount Zalmon, he and all the people who were with him. And Abimelek took an ax in his hand and cut a bundle of wood and lifted it and put it on his shoulder and said to the people with him, "What you've seen me do: hurry and do like me." 49. And all of the people, likewise, each cut a bundle and followed Abimelek and put them at the stronghold and set the stronghold on fire by them. And all the people of Shechem Tower died as well, about a thousand men and women.

50. And Abimelek went to Thebez and encamped against Thebez and captured it. 51. And there was a strong tower inside

the city, and all the men and the women and all the city's citizens fled there and closed themselves in, and they went up on the roof of the tower. 52. And Abimelek came to the tower and fought against it, and he went up to the tower's entrance to burn it with fire. 53. And a woman threw an upper millstone down on Abimelek's head and crushed his skull. 54. And he called quickly to the young man, his armor-bearer, and said to him, "Draw your sword and kill me, or else they'll say about me, 'A woman killed him.'" And his young man ran him through, and he died. 55. And the men of Israel saw that Abimelek was dead, and they each went to his place. 56. And so God brought back Abimelek's wrongdoing which he had done to his father by killing his seventy brothers, 57. and God brought back on their heads all the wrongdoing of the people of Shechem; and the curse of Jerubbaal's son, Jotham, came to them.

10:8. And [the children of Ammon] crashed and crushed the children of Israel in that year, for eighteen years, all the children of Israel who were beyond the Jordan, in the land of the Amorite that was in Gilead. 9. And the children of Ammon crossed the Jordan to fight Judah and Benjamin and the house of Ephraim as well, and Israel was very distressed. 17. And the children of Ammon were called up, and they camped in Gilead, and the children of Israel gathered and camped in Mizpah. 18. And the people, the officers of Gilead, said to one another, "Who's the man who'll set out to fight the children of Ammon? He'll become the head of everyone who lives in Gilead."

11:1. And Jephthah, the Gileadite, was a powerful warrior, and he was a son of a prostitute woman. And Gilead fathered Jephthah. 2. And Gilead's wife gave birth to sons by him, and the wife's sons grew up, and they expelled Jephthah and said to him, "You won't inherit in our father's house, because you're a son of another woman." 3. And Jephthah fled from his brothers and lived in the land of Tob, and worthless people collected around Jephthah and went out with him.

4. And it was, as the days passed, that the children of Ammon fought with Israel. 5. And it was when the children of Ammon fought with Israel: and Gilead's elders went to get Jephthah from the land of Tob, 6. and they said to Jephthah, "Come, and you'll be our commander, and let's fight with the children of Ammon."

7. And Jephthah said to Gilead's elders, "Haven't *you* hated me and driven me from my father's house? And why have you come to me now when you have distress?"

8. And Gilead's elders said to Jephthah, "That's why we've come back to you now, and if you'll come with us and fight against the children of Ammon then you'll be our head of everyone who lives in Gilead."

9. And Jephthah said to Gilead's elders, "If you bring me back to fight against the children of Ammon, and YHWH delivers them before me, I'll become your head."

10. And Gilead's elders said to Jephthah, "YHWH will be listening between us that we'll do according to your word."

11. And Jephthah went with Gilead's elders, and the people made him a head and commander over them. And Jephthah spoke all his words before YHWH in Mizpah.

12. And Jephthah sent messengers to the king of the children of Ammon, saying, "What business do I and you have that you've come to me to fight in my land?"

13. And the king of the children of Ammon said to Jephthah's messengers, "Because Israel took my land when it came up from Egypt, from Arnon to the Jabbok to the Jordan. And now, give them back peaceably."

14. And Jephthah went further and sent messengers to the king of the children of Ammon 15. and said to him, "Jephthah says this: Israel didn't take the land of Moab and the land of the children of Ammon. 16. But when they came up from Egypt, and Israel went in the wilderness to the Red Sea and came to Kadesh, 17. Israel sent messengers to the king of Edom saying, 'Let me pass through your land,' and the king of Edom didn't listen. And he sent to the king of Moab as well, and he wasn't willing. And Israel lived in Kadesh. 18. And they went in the wilderness and went around the land of Edom and the land of Moab and came from the east of the land of Moab and camped beyond Arnon and didn't come inside Moab's border, because Arnon is Moab's border. 19. And Israel sent messengers to Sihon, king of the Amorite, king of Heshbon, and Israel said to him, 'Let us pass through your land to my place.' 20. And Sihon didn't trust Israel to pass his border, and Sihon gathered all his people and camped in Jahaz and fought with Israel. 21. And YHWH, Israel's God, delivered Sihon and all his people into Israel's hand, and they struck them, and Israel possessed all the land of the Amorite, who lived in that land, 22. and they possessed all the Amorite border, from Arnon to the Jabbok and from the wilderness to the Jordan. 23. And now YHWH, Israel's God, has dispossessed the Amorite from before his people, Israel. And will *you* possess it?! 24. Is it not that whatever Chemosh, your god, causes you to possess, you'll possess *that?!* And we'll possess everything that YHWH, our God, has dispossessed before us. 25. And now, are you any better than Balak, son of Zippor, king of Moab? Did he *quarrel* with Israel or *fight* against them? 26. When Israel lived in Heshbon and in its environs and in Aroer and in its environs and in all the cities that were near Arnon, three hundred years, why didn't you recover them at that time? 27. And I haven't sinned against you, and you do me

wrong to fight against me. May YHWH, the judge, judge today between the children of Israel and the children of Ammon!"

28. And the king of the children of Ammon did not listen to Jephthah's words that he had sent to him.

29. And YHWH's spirit was on Jephthah, and he passed through Gilead and Manasseh and passed through Mizpeh Gilead, and from Mizpeh Gilead he passed to the children of Ammon. 30. And Jephthah made a vow to YHWH and said, "If you'll *deliver* the children of Ammon into my hand, 31. then whatever will come out of the doors of my house to me when I come back in peace from the children of Ammon will be YHWH's, and I'll offer it up as a burnt offering!"

32. And Jephthah passed to the children of Ammon to fight against them, and YHWH delivered them into his hand, 33. and he struck them from Aroer to where you come to Minnith, twenty cities, and to Abel Keramim, a very big blow. And the children of Ammon were subdued before the children of Israel.

34. And Jephthah came to Mizpah, to his house, and here was his daughter coming out to him with tambourines and with dances. And she was an only child. He had no other son or daughter. 35. And it was, when he saw her, that he ripped his clothes and said, "Oh, my daughter, you've *brought me down,* and *you're* among the ones who anguish me! And *I've* opened my mouth to YHWH and can't go back!"

36. And she said to him, "My father, you've opened your mouth to YHWH. Do to me whatever came out of your mouth, because YHWH has made retributions for you from your enemies, from the children of Ammon." 37. And she said to her father, "Let this thing be done for me: Hold off from me for two months, and let me go, and I'll go down along the mountains, and I'll weep over my virginity, I and my companions."

38. And he said, "Go."

And he sent her off for two months, and she went, she and her companions, and she wept over her virginity on the mountains. 39. And it was at the end of two months, and she came back to her father, and he carried out his vow that he had made on her. And she had never known a man. And it became a custom in Israel: 40. the daughters of Israel would go regularly to lament the daughter of Jephthah the Gileadite, four days in the year.

13:2. And there was a man from Zorah, from a Danite family, and his name was Manoah, and his wife was barren. She had never given birth. 3. And an angel of YHWH appeared to the woman and said to her, "Here, you're barren, and you've never given birth, and you'll become pregnant and give birth to a son. 4. And now watch yourself and don't drink wine and beer and don't eat anything unclean 5. because, here, you'll become pregnant, and you'll give birth to a son, and a razor shall not go up on his head, because the boy will be a nazirite of God from the womb, and *he'll* begin to save Israel from the Philistines' hand!"

6. And the woman came and said to her husband, saying, "A man of God came to me. And his appearance was like the appearance of an angel of God: very awesome. And I didn't ask him where he was from, and he didn't tell me his name. 7. And he said to me, 'Here, you'll become pregnant, and you'll give birth to a son. And now don't drink wine and beer and don't eat anything unclean, because the boy will be a nazirite of God from the womb to the day of his death.'"

8. And Manoah prayed to YHWH and said, "Please, my Lord, let the man of God whom you sent come to us again and instruct us what we should do for the boy who'll be born."

9. And God listened to Manoah's voice, and the angel of God came to the woman again. And she was sitting in the field, and

Manoah, her husband, was not with her. 10. And the woman hurried and ran and told her husband. And she said to him, "Here, the man who came to me that day has appeared to me."

11. And Manoah got up and followed his wife. And he came to the man and said to him, "Are you the man who spoke to the woman?"

And he said, "I am."

12. And Manoah said, "Now let your words come to pass. What will the boy's rule be, and his practice?"

13. And the angel of YHWH said to Manoah, "Watch yourself to do everything that I said to your wife. 14. She shall not eat anything that comes from the wine vine, and let her not drink wine and beer, and let her not eat anything unclean. She shall be watchful of everything that I've commanded her."

15. And Manoah said to the angel of YHWH, "Let us detain you, and we'll make a goat kid for you."

16. And the angel of YHWH said to Manoah, "If you'll detain me, I won't eat your bread. But if you'll make a burnt offering, offer it up to YHWH"—because Manoah did not know that it was an angel of YHWH.

17. And Manoah said to the angel of YHWH, "What's your name—so that when your word comes to pass we'll honor you."

18. And the angel of YHWH said to him, "Why is this that you ask my name? And it is my wonder."

19. And Manoah took the goat kid and the grain offering and offered them on a rock to YHWH. And he did a wondrous thing, and Manoah and his wife were seeing it—20. and it was as the flame was going up from the altar to the skies: and the angel of YHWH went up in the flame of the altar. And Manoah and his wife had seen it, and they fell on their faces to the ground. 21. And the angel of YHWH did not go on to appear again to Manoah and

his wife. Then Manoah knew that it was an angel of YHWH. 22. And Manoah said to his wife, "We're going to *die,* because we've seen God!"

23. And his wife said to him, "If YHWH desired to kill us He wouldn't have taken a burnt offering and a grain offering from our hand, and He wouldn't have shown us all these things and wouldn't have had us hear such things at this time."

24. And the woman gave birth to a son and called his name Samson. And the boy grew, and YHWH blessed him. 25. And YHWH's spirit began to move him in the camp of Dan, between Zorah and Eshtaol.

14:1. And Samson went down to Timnah and saw a woman in Timnah of the Philistines' daughters, 2. and he went up and told his father and his mother and said, "I've seen a woman in Timnah of the Philistines' daughters. And now, get her for me as a wife."

3. And his father and his mother said to him, "Is there no woman among your brothers' daughters and among all your people, that you go to get a wife from the uncircumcised Philistines?!"

And Samson said to his father, "Get her for me, because she's right in my eyes."

4. And his father and his mother did not know that it was from YHWH, because He was seeking an opportunity from the Philistines, for the Philistines were dominating Israel at that time. 5. And Samson and his father and his mother went down to Timnah and came to the vineyards of Timnah. And here was a lion roaring at him, 6. and YHWH's spirit rushed onto him, and he tore it apart like tearing apart a kid, and he had nothing in his hand. And he did not tell his father and his mother what he had done. 7. And he went down and spoke to the woman, and she was right in Samson's eyes. 8. And he came back to get her after some time, and he turned to see the remains of the lion, and here was a swarm of bees in the

lion's carcass—and honey. 9. And he scraped it into his hands and went, eating as he went. And he went to his father and to his mother and gave them some, and they ate, and he did not tell them that he had scraped the honey from the lion's carcass. 10. And his father went down to the woman. And Samson made a feast there because that is what the young men would do. 11. And it was, when they saw him, that they took thirty companions, and they were with him, 12. and Samson said to them, "Let me give you a riddle. If you can *explain* it for me in the seven days of the feast and find it out, then I'll give you thirty capes and thirty changes of clothing. 13. And if you won't be able to explain it for me, then *you'll* give me thirty capes and thirty changes of clothing."

And they said to him, "Give us your riddle and let's hear it."

14. And he said to them,

> From the one who feeds: food came out
> And from the strong: sweet came out.

And they were not able to explain the riddle for three days. 15. And it was on the seventh day, and they said to Samson's wife, "Trick your husband so he'll explain the riddle for us or else we'll burn you and your father's house in fire. Did you invite us here to dispossess us?!"

16. And Samson's wife wept upon him and said, "You just hate me, and you don't love me. You've given the riddle to my countrymen, and you haven't explained it to me."

And he said to her, "Here, I haven't explained it to my father and to my mother, and shall I explain it to you?"

17. And she wept upon him for the seven days that they had the feast. And it was on the seventh day, and he explained it to her because she pressured him. And she explained the riddle to her

countrymen. 18. And the people of the city said to him on the seventh day before the sun went down,

> What is sweeter than honey?
> And what is stronger than a lion?

And he said to them, "If you hadn't plowed with my heifer you wouldn't have found out my riddle!" 19. And YHWH's spirit rushed onto him, and he went down to Ashkelon and struck thirty of their men and took what could be pulled off them and gave the changes of clothing to those who explained his riddle. And his anger flared, and he went up to his father's house. 20. And Samson's wife became the wife of one of his companions—who had been his companion.

15:1. And it was after some time, in the days of the wheat harvest, and Samson visited his wife with a goat kid and said, "Let me come to my wife in the room."

And her father would not let him come. 2. And her father said, "I'd *said* that you *hated* her, and I gave her to your companion. Isn't her younger sister better than she is? Let her be yours instead of her."

3. And Samson said to them, "This time I'm free of blame from the Philistines when I do them wrong." 4. And Samson went and captured three hundred foxes and took torches; and he turned them tail to tail and put one torch between the two tails, in the middle, 5. and he set the torches on fire and let them go in the Philistines' standing grain, and it burned from the shocks to the standing grain and to the vineyard and olive tree.

6. And the Philistines said, "Who did this?"

And they said, "Samson, the Timnite's son-in-law, because he took his wife and gave her to his companion." And the Philistines went up and burned her and her father in fire.

7. And Samson said to them, "If you do a thing like this, I'll certainly have revenge against you—and after that I'll hold back." 8. And he struck them, hip on thigh, a big blow. And he went down and lived at the cleft of the rock of Etam.

9. And the Philistines went up and camped in Judah and were spread out in Lehi. 10. And the men of Judah said, "Why have you come up against us?"

And they said, "We've come up to tie up Samson, to do to him as he did to us."

11. And three thousand men from Judah went down to the cleft of the rock Etam and said to Samson, "Didn't you know that the Philistines dominate us? And what is this you've done to us?"

And he said to them, "As they did to me, that's what I did to them."

12. And they said to him, "We've come down to tie you up, to deliver you into the Philistines' hand."

And Samson said to them, "Swear to me that you won't injure me yourselves."

13. And they said to him, saying, "No, but we'll *tie you up* and deliver you into their hand, and we won't *kill* you."

And they tied him up with two new ropes and brought him up from the rock. 14. He had come to Lehi, and the Philistines had shouted at him, and YHWH's spirit rushed onto him, and the ropes that were on his arms were like flax that burned in fire, and his restraints melted from his hands. 15. And he found a fresh jawbone of an ass and put out his hand and took it and struck a thousand men with it. 16. And Samson said,

> With an ass's jawbone
> heap of heaps,

with an ass's jawbone
I struck a thousand men.

17. And it was as he finished speaking, and he threw the jawbone down from his hand. And he called that place Ramat-Lehi. 18. And he was very thirsty, and he called to YHWH and said, "You delivered this big salvation into your servant's hand, and now I'll die of thirst and fall into the hand of the uncircumcised!" 19. And God split open a hollow place that was in Lehi, and water came out of it, and he drank, and his spirit came back, and he lived. On account of this he called its name En-hakkore, which is in Lehi to this day.

16:1. And Samson went to Gaza and saw a prostitute there and came to her. 2. And it was told to the Gazites, saying, "Samson has come here." And they went around and made an ambush for him all night at the city's gate, and they kept quiet all night, saying, "Until the morning light, and we'll kill him."

3. And Samson lay until midnight, and he got up at midnight and took hold of the doors of the city's gate and the two doorposts and pulled them up along with the bolt and put them on his shoulders and took them up to the top of the mountain that faces Hebron.

4. And it was after this, and he loved a woman at the wadi Sorek, and her name was Delilah. 5. And the lords of the Philistines went up to her and said to her, "Trick him and see how he gets his big strength and how we can prevail against him, and we'll tie him up to degrade him. And we'll each give you eleven hundred weights of silver."

6. And Delilah said to Samson, "Tell me, how do you get your big strength? And how could you be tied up to degrade you?"

7. And Samson said to her, "If they'd tie me up with seven fresh cords that haven't been dried, then I'd weaken and be like any man."

8. And the lords of the Philistines brought up seven fresh cords that had not been dried to her, and she tied him up with them. 9. And the ambush was ready for her in the room, and she said to him, "The Philistines are on you, Samson!" And he broke the cords the way a string of tow is broken when fire touches it. And his strength's source was not made known.

10. And Delilah said to Samson, "Here, you've toyed with me and spoken lies to me! Now tell me, how could you be tied up?"

11. And he said to her, "If they'd *tie me up* with new ropes, with which work has never been done, then I'd weaken and be like any man."

12. And Delilah took new ropes and tied him up with them and said to him, "The Philistines are on you, Samson!" And the ambush was ready in the room. And he broke them from his arms like a thread.

13. And Delilah said to Samson, "You've toyed with me and spoken lies to me this far. Tell me, how could you be tied up?"

And he said to her, "If you'd weave seven locks of my head into the web and fasten it with the pin into the wall, then I'd weaken and be like any man."

14. And it was when he was asleep: and Delilah took seven locks of his head and wove them into the web and fastened it with the pin and said to him, "The Philistines are on you, Samson!" And he woke from his sleep and pulled up the pin, the loom, and the web.

15. And she said to him, "How can you say, 'I love you,' when your heart isn't with me. Three times now you've toyed with me and haven't told me how you get your big strength."

16. And it was because she pressured him with her words every day and prodded him, and his soul was wearied to death, 17. that he told her all his heart and said to her, "A razor hasn't gone up on my head because I'm a nazirite of God from my mother's womb. If I

were shaved, then my strength would turn away from me and I'd weaken and be like any man."

18. And Delilah saw that he had told her all his heart, and she sent and called to the lords of the Philistines, saying, "Come up this time, because he's told me all his heart." And the lords of the Philistines came up to her, and they came up with the silver in their hand. 19. And she had him sleep on her knees, and she called a man and shaved the seven locks of his head, and she began to degrade him, and his strength turned away from him. 20. And she said, "The Philistines are on you, Samson!"

And he woke from his sleep and said, "I'll get out and shake free like every other time," and he did not know that YHWH had turned away from him. 21. And the Philistines took hold of him and gouged his eyes and brought him down to Gaza and bound him with bronze fetters, and he was grinding at the mill in the prison house.

22. And the hair of his head began to grow where it had been cut.

23. And the lords of the Philistines were gathered to make a big sacrifice to Dagon, their god, and for festivity. And they said, "Our god has delivered Samson, our enemy, into our hand."

24. And the people saw him, and they praised their god, because they said, "Our god has delivered into our hand our enemy and the destroyer of our land and the one who has made our slain so many." 25. And it was when their heart was in a good mood: and they said, "Call Samson and let him perform for us!" And they called Samson from the prison house, and he performed in front of them. And they stood him between the columns.

26. And Samson said to the boy who was holding his hand, "Leave me and let me feel the columns on which the building is constructed, and I'll lean on them." 27. And the building was filled with men and women, and all the lords of the Philistines were

there, and about three thousand men and women who were watching as Samson performed were on the roof. 28. And Samson called to YHWH and said, "My Lord, YHWH, remember me; and give me strength just this one time, God, and let me have revenge from the Philistines for one of my two eyes." 29. And Samson took hold of the two middle pillars on which the building was constructed and pressed on them, one with his right and one with his left. 30. And Samson said, "Let my soul die with the Philistines." And he pushed with strength, and the building fell on the lords and on all the people who were in it. And the dead whom he killed in his death were more than those he had killed in his life.

31a. And his brothers and all of his father's house went down and lifted him and came up and buried him between Zorah and Eshtaol, in his father Manoah's tomb.

17:1 And there was a man from the mountains of Ephraim, and his name was Micah. 2. And he said to his mother, "The one thousand, one hundred weights of silver that were taken from you—and you cursed, and you said it in my ears, too: here's the silver with me; I took it."

And his mother said, "My son is blessed to YHWH." 3. And he gave back the one thousand, one hundred weights of silver. And his mother said, "I've consecrated the silver to YHWH, from my hand to my son's, to make a graven image and a molten image. And now I give it back to you."

4. And he gave back the silver to his mother, and his mother took two hundred weights of silver and gave it to the metalworker, and he made it into a graven image and a molten image. And it was in Micah's house. 5. And the man Micah had a house of God, and

he made an ephod and teraphim, and he designated one of his sons, and he became his priest.

6. In those days there was no king in Israel. Each person did what was right in his eyes.

7. And there was a young man from Bethlehem, Judah, from the family of Judah, and he was a Levite, and he lived there. 8. And the man went from the city—from Bethlehem, Judah—to live in whatever place he would find, and he came to the mountains of Ephraim, to Micah's house, making his way. 9. And Micah said to him, "Where are you coming from?"

And he said to him, "I'm a Levite from Bethlehem, Judah, and I'm going to live in whatever place I find."

10. And Micah said to him, "Live with me and become a father and priest to me, and I'll give you ten weights of silver per year and a set of clothes and your living." And the Levite came. 11. And the Levite was content to live with the man, and the young man was like one of his sons to him. 12. And Micah designated the Levite, and the young man became his priest, and he was in Micah's house. 13. And Micah said, "Now I know that YHWH will be good to me, because I have the Levite as a priest."

18:1. In those days there was no king in Israel. And in those days the Danite tribe was seeking a place to live as its legacy because none had fallen to it among the tribes of Israel up to that day. 2. And the children of Dan sent five men from their family, out of all of them, men who were warriors, from Zorah and from Eshtaol, to spy out the land and to explore it. And they said to them, "Go, explore the land." And they came to the mountains of Ephraim, to Micah's house, and spent the night there.

3. They were at Micah's house, and they recognized the young Levite man's accent, and they turned there and said to him, "Who

brought you here? And what are you doing in this place? And what do you have here?"

4. And he said to them, "Micah did such and such for me, and he hired me, and I became his priest."

5. And they said to him, "Ask God so we may know: will our way that we're going on be successful?"

6. And the priest said to them, "Go in peace. The way in which you're going is before YHWH." 7. And the five of them went. And they came to Laish and saw the people who were in it, living securely in the Sidonians' fashion: quiet and secure, and without anyone to injure them in anything in the land, possessing wealth; and they were far from the Sidonians and had no business with anyone.

8. And they came to their brothers at Zorah and Eshtaol, and their brothers said to them, "What do you say?"

9. And they said, "Get up, and let's go up against them, because we've seen the land, and, here, it's very good, and you're keeping quiet! Don't hesitate to go—to come to possess the land. 10. When you come, you'll come to a secure people, and the land is wide open because God has delivered it into your hand, a place where there's no lack of anything that's in the earth." 11. And six hundred men from the Danite family, from Zorah and Eshtaol, traveled from there, armed with weapons of war. 12. And they went up and camped in Kiriath-jearim in Judah. (On account of this they've called that place "Camp of Dan" to this day. Here, it's behind Kiriath-jearim.) 13. And they passed from there to the mountains of Ephraim and came to Micah's house.

14. And the five men who had gone to spy out the land of Laish answered and said to their brothers, "Did you know that there are an ephod and teraphim and a graven image and a molten image in these houses? And now you know what you're going to do!" 15.

And they turned from there and came to the Levite young man's house, at Micah's house, and asked him how he was. 16. And six hundred men who were from the children of Dan, armed with their weapons of war, were standing at the entrance of the gate. 17. And the five men who had gone to spy out the land came up, they came there, they took the graven image and the ephod and the teraphim and the molten image. And the priest stood at the entrance of the gate, and the six hundred men who were armed with weapons of war, 18. and these came to Micah's house. And they took the graven image, the ephod, and the teraphim and the molten image.

And the priest said to them, "What are you doing?"

19. And they said to him, "Keep quiet! Put your hand over your mouth! And go with us and become our father and priest. Is it better for you to be a priest for one man's house or for you to be a priest for a tribe and family in Israel?!" 20. And the priest's heart was glad. And he took the ephod and the teraphim and the graven image and came among the people, 21. and they turned and went. And they put the infants and the cattle and the goods ahead of them. 22. They went far from Micah's house, and Micah and the people who were in the houses that were with Micah's house rallied, and they caught up to the children of Dan. 23. And they called to the children of Dan, and they turned their faces and said to Micah, "What business do you have that you've rallied?"

24. And he said, "You've taken my gods that I made and the priest, and you've gone. And what else do I have? And what is this you say to me: 'What business do you have?'!"

25. And the children of Dan said to him, "Don't let your voice be heard among us, or else fierce-souled people will injure you, and you'll lose your life and the lives of your house." 26. And the children of Dan went their way. And Micah saw that they were

stronger than he, and he turned and went back to his house. 27. And they took what Micah had made and the priest who had been his.

And they came upon Laish, upon a people quiet and secure, and they struck them with the edge of the sword and burned the city in fire, 28. and there was none to rescue because it was far from Sidon, and they had no business with anyone. And it was in the valley that belonged to Beth-Rehob, and they built the city and lived in it. 29. And they called the name of the city Dan, after the name of their father Dan who was born to Israel, though in fact Laish was the name of the city at first.

30. And the children of Dan set up the graven image for themselves. And Jonathan, son of Gershom, son of Moses: he and his sons were priests for the Danite tribe. 31. And they maintained for themselves Micah's graven image that he had made, all the days that the House of God was in Shiloh.

19:1. And it was in those days, and there was no king in Israel. And there was a Levite man living on the slopes of the mountains of Ephraim, and he took himself a concubine woman from Bethlehem, Judah. 2. And his concubine whored and went from him to her father's house, to Bethlehem, Judah, and she was there for a period of four months. 3. And her man got up and went after her to speak on her heart to bring her back—and his boy was with him and a pair of asses—and she brought him to her father's house, and the young woman's father saw him and was happy to meet him. 4. And his father-in-law, the young woman's father, held him, and he lived with him three days, and they ate and drank and stayed the night there.

5. And it was on the fourth day, and they got up early in the morning and got up to go. And the young woman's father said to his son-in-law, "Satisfy your heart with a piece of bread, and afterwards you'll go." 6. And the two of them sat and ate together, and drank. And the young woman's father said to the man, "Be content and spend the night and let your heart be in a good mood." 7. And the man got up to go, and his father-in-law pressed him, and he came back and spent the night there.

8. And he got up early in the morning on the fifth day to go, and the young woman's father said, "Satisfy your heart." And they delayed until the day stretched on, and the two of them ate. 9. And the man got up to go, he and his concubine and his boy. And his father-in-law, the young woman's father, said to him, "Here, the day is waning toward evening. Spend the night. Here, the day is drawing to a close. Spend the night here, and let your heart be in a good mood, and you'll get up early tomorrow to be on your way, and you'll go to your tent." 10. And the man was not willing to spend the night, and he got up and went. And he came to the area adjacent to Jebus, which is Jerusalem, and a pair of saddled asses were with him, and his concubine was with him. 11. They were by Jebus, and the day was very far along.

And his boy said to his lord, "Come on, and let's turn to this city of the Jebusite and spend the night in it."

12. And his lord said to him, "We won't turn to a city of a foreigner, who's not of the children of Israel, here, but we'll pass on to Gibeah." 13. And he said to his boy, "Come, and let's go near to one of the places and spend the night in Gibeah or in Ramah." 14. And they passed on and went. And the sun set on them near Gibeah, which is part of Benjamin, 15. and they turned there to come to spend the night in Gibeah. And he came and sat in the city

square, and no one was taking them home to spend the night. 16. And here was an old man coming from his affairs, from the field, in the evening—and the man was from the mountains of Ephraim, and he was residing in Gibeah, while the people of the place were Benjaminites. 17. And he raised his eyes and saw the man who was passing through in the city square.

And the old man said, "Where are you going, and where are you coming from?"

18. And he said to him, "We're passing from Bethlehem, Judah, to the slopes of the mountains of Ephraim. I'm from there, and I went to Bethlehem, Judah, and I'm going to the House of YHWH. And no one is taking me home, 19. and our asses have both straw and fodder, and I and your maid and the boy have bread and wine with your servants. There's no lack of anything."

20. And the old man said, "Peace to you. Only all of your needs are on me. Only don't spend the night in the square." 21. And he had him come to his house and he fed the asses, and they washed their feet and ate and drank.

22. They were getting their hearts in a good mood, and, here, people of the city, good-for-nothing people, surrounded the house, pounding on the door. And they said to the old man who was the owner of the house, saying, "Bring out the man who came to your house, and let's know him!"

23. And the man who was the owner of the house went out to them and said to them, "Don't, my brothers. Don't do bad. Since this man came to my house, don't do this foolhardy thing. 24. Here are my virgin daughter and his concubine. Let me bring them out. And degrade them and do to them what's good in your eyes, and don't do this foolhardy thing to this man."

25. And the people were not willing to listen to him. And the man took hold of his concubine and brought her to them outside. And

they knew her and abused her all night until morning and let her go at sunrise. 26. And the woman came toward morning and fell at the entrance of the house where her lord was, until it was light. 27. And her lord got up in the morning and opened the doors of the house and went out to go on his way, and here was the woman, his concubine, fallen at the entrance of the house, and her hands on the threshold.

28. And he said to her, "Get up, and let's go."

And there was no answer.

And he took her on the ass, and the man got up and went to his place. 29. And he came to his house and took a knife and took hold of his concubine and cut her up, down to her bones, into twelve parts, and sent her in all of Israel's borders. 30. And everyone who saw it said, "Nothing like this has happened or been seen from the day the children of Israel went up from the land of Egypt until this day. Consider it, take counsel, and speak!" 20:1. And all the children of Israel went out, and the congregation assembled as one man, from Dan to Beer-sheba and the land of Gilead, to YHWH at Mizpeh. 2. And the chiefs of all the people, all the tribes of Israel, stood in the assembly of the people of God, four hundred thousand sword-drawing men on foot.

3. And the children of Benjamin heard that the children of Israel had gone up to Mizpeh.

And the children of Israel said, "Speak. How did this bad thing happen?"

4. And the Levite man, the murdered woman's man answered, and he said, "I and my concubine came to Gibeah, which is part of Benjamin, to spend the night. 5. And the citizens of Gibeah came up against me and surrounded the house against me at night. They meant to kill me, and they degraded my concubine, and she died. 6. And I took hold of my concubine and cut her up and sent her in all the territory of Israel's legacy because they did a wicked thing and

a foolhardy thing in Israel. 7. Here you're all children of Israel. Give word and counsel here."

8. And all the people got up as one man, saying, "We won't each go to his tent, and we won't each turn to his house. 9. And now this is the thing that we'll do to Gibeah: We'll go against it by lot, 10. and we'll take ten people per hundred for all the tribes of Israel, and a hundred per thousand, and a thousand per ten thousand, to take provisions for the people, to do when they come to Gibeah, Benjamin, according to all the foolhardy thing that it did in Israel." 11. And all the men of Israel gathered to the city, united, like one man.

12. And the tribes of Israel sent people through all the tribe of Benjamin, saying, "What is this bad thing that has happened among you? 13. And now give the good-for-nothing people who are in Gibeah, and we'll kill them and burn out the bad from Israel."

And the children of Benjamin were not willing to listen to the voice of their brothers, the children of Israel. 14. And the children of Benjamin gathered from the cities to Gibeah to go out to war against the children of Israel. 15. And the children of Benjamin were mustered in that day from the cities: twenty-six thousand sword-drawing men, outside of the those who lived in Gibeah, seven hundred chosen men. 16. Out of all this people, seven hundred chosen men bound their right hand, every one of whom could sling a stone at a hair and not miss.

17. And the men of Israel were mustered, outside of Benjamin: four hundred thousand sword-drawing men, every one of whom was a man of war. 18. And they got up and went up to Beth-El and asked of God, and the children of Israel said, "Who will go up first for us for the war against the children of Benjamin?"

And YHWH said, "Judah first."

19. And the children of Israel got up in the morning and camped against Gibeah. 20. And the men of Israel went out to the war with

Benjamin, and the men of Israel aligned with them for war toward Gibeah. 21. And the children of Benjamin came out of Gibeah and destroyed twenty-two thousand men of Israel to the ground in that day. 22. And the people, the men of Israel, fortified themselves and aligned again for war in the place where they had aligned on the first day. 23. And the children of Israel went up and wept before YHWH until the evening and asked of YHWH, saying, "Shall I approach again for war with the children of Benjamin, my brother?"

And YHWH said, "Go up against him."

24. And the children of Israel came near to the children of Benjamin on the second day. 25. And Benjamin came out toward them from Gibeah on the second day and destroyed another eighteen thousand men of the children of Israel to the ground; all of these were sword-drawing. 26. And all the children of Israel and all the people went up and came to Beth-El and wept and sat there before YHWH and fasted in that day until evening, and they made burnt offerings and peace offerings before YHWH. 27. And the children of Israel asked of YHWH—and the ark of God's covenant was there in those days; 28. and Phinehas, son of Eleazar, son of Aaron, was standing in front of it in those days—saying, "Shall I go out again for war with the children of Benjamin, my brother, or shall I stop?"

And YHWH said, "Go up, because tomorrow I'll deliver them into your hand."

29. And Israel set ambushers around Gibeah.

30. And the children of Israel went up to the children of Benjamin on the third day and aligned toward Gibeah as at the other times. 31. And the children of Benjamin went out toward the people. They were drawn away from the city. And they began to strike some of the people—dead bodies—as at the other times, in the

highways, one of which goes up to Beth-El and one to Gibeah, in the field: about thirty men of Israel.

32. And the children of Benjamin said, "They're routed in front of us like before."

And the children of Israel said, "Let's flee and draw them away from the city to the highways." 33. And every man of Israel got up from his place, and they aligned at Baal-Tamar, and Israel's ambush burst out from its place, from Maareh-Geba, 34. and ten thousand men chosen out of all Israel came from opposite Gibeah. And the war was heavy. And they did not know that the bad was overtaking them. 35. And YHWH struck Benjamin before Israel, and the children of Israel destroyed twenty-five thousand, one hundred men from Benjamin on that day, all of these sword-drawing. 36. And the children of Benjamin saw that they were routed. And the men of Israel gave ground for Benjamin because they relied on the ambush that they had set at Gibeah. 37. And the ambushers hurried and charged at Gibeah, and the ambushers proceeded and struck the whole city with the edge of the sword. 38. And there was an appointed signal between the men of Israel and the ambushers: to send up a great stream of smoke from the city, 39. and the men of Israel would reverse in the battle. And Benjamin had begun to strike, dead bodies among the men of Israel, about thirty men, because they had said, "He's *routed* before us, like the first battle." 40. And the stream, a column of smoke, had begun to go up from the city. And Benjamin turned back, and here was the whole city going up to the skies. 41. And the men of Israel reversed, and the men of Benjamin were terrified because they saw that the bad had overtaken them, 42. and they turned before the men of Israel to the way to the wilderness, but the war stuck with them. And they were destroying those from the city inside it. 43. They surrounded Benjamin. They pursued them. They trampled them easily as far as opposite Gibeah toward the sunrise. 44. And eighteen thousand men from Benjamin fell. All of these were warriors. 45. And they turned and fled to the wilderness, to the rock of Rimmon, and five thousand men were cut down in the highways. And they stuck with them as far as Gidom and struck two thousand men from them. 46. And all the fallen from Benjamin were twenty-five thousand sword-drawing men on that day. All of these were warriors. 47. And six hundred men turned and fled to the wilderness, to the rock of Rimmon, and lived at the rock of Rimmon four months. 48. And the people of Israel went back to the children of Benjamin and struck them from each city by the edge of the sword: man to beast to everything that was found. They set fire to all the cities that were found as well.

21:1. And the men of Israel had sworn at Mizpeh, saying, "No man among us will give his daughter to Benjamin as a wife." 2. And the people came to Beth-El and sat there until evening before God and raised their voices and wept, a big grieving, 3. and said, "Why, YHWH, God of Israel, has this happened in Israel, that one tribe should be eliminated from Israel?" 4. And it was the following day, and the people got up early and built an altar there and made burnt offerings and peace offerings. 5. And the children of Israel said, "Who is there, of all the tribes of Israel, who didn't come up to YHWH in the assembly?" Because the big oath was regarding whoever did not go up to YHWH at Mizpeh, saying, "He shall be *put to death.*" 6. And the children of Israel had regret toward Benjamin, their brother; and they said, "One tribe has been cut off from Israel today. 7. What shall we do for them, for those who are left, for wives, when we've sworn by YHWH not to give any of our daughters to them as wives?" 8. And they said, "What one is there of the tribes of Israel who didn't go up to YHWH at Mizpeh?" And, here, not a man had come to the camp from

Jabesh-Gilead to the assembly. 9. And the people were mustered; and, here, there was not a man there from those who lived in Jabesh-Gilead. 10. And the congregation sent twelve thousand men of the warriors there, and they commanded them, saying, "Go and strike those who live in Jabesh-Gilead by the sword, and the women and the infants. 11. And this is what you shall do: you shall completely destroy every male and every woman who has known lying with a male." 12. And they found among those who lived in Jabesh-Gilead four hundred virgin young women who had not known a man—lying with a male—and they brought them to the camp at Shiloh, which is in the land of Canaan.

13. And all the congregation sent and spoke to the children of Benjamin who were at the rock of Rimmon and offered them peace. 14. And Benjamin came back at that time, and they gave them the women whom they had kept alive of the women of Jabesh-Gilead, and they did not find enough for them. 15. And the people were reconciled toward Benjamin, for YHWH had made a breach among the tribes of Israel.

16. And the elders of the congregation said, "What shall we do for women for those who are left since woman is destroyed from Benjamin?" 17. And they said, "There must be a possession for Benjamin's survivors so a tribe won't be wiped out from Israel. 18. And *we're* not able to give them wives from our daughters because the children of Israel swore, saying, 'Cursed is anyone who gives a woman to Benjamin.'"

19. And they said, "Here's a holiday of YHWH in Shiloh regularly which is north of Beth-El, east of the highway that goes up from Beth-El to Shechem, and south of Lebonah." 20. And they commanded the children of Benjamin, saying, "Go and make an ambush in the vineyards. 21. And you'll look; and, here, if the

daughters of Shiloh go out to do dances, then you'll come out of the vineyards, and each of you shall catch his woman from the daughters of Shiloh and go to the land of Benjamin. 22. And it will be, when their fathers or their brothers come to us to protest, that we'll say to them, 'Be gracious to them for us, because each of us didn't take his woman in the war, and because you didn't give them: you'd be guilty by now.'"

23. And the children of Benjamin did this, and they carried off women according to their number whom they seized from those who were dancing. And they went and came back to their territory and rebuilt the cities and lived in them. 24. And each of the children of Israel went from there at that time to his tribe and to his family, and each man went from there to his territory.

25. In those days there was no king in Israel. Each man would do what was right in his eyes.

1 Samuel 1:1. And there was a man from Ramathaim Zophim, from the mountains of Ephraim, and his name was Elkanah, son of Jeroham, son of Elihu, son of Tohu, son of Zuph, an Ephraimite. 2. And he had two wives. One's name was Hannah, and the second's name was Peninnah. And Peninnah had children, and Hannah did not have children. 3. And that man went up from his city regularly to bow and to sacrifice to YHWH of hosts in Shiloh. And Eli's two sons, Hophni and Phinehas, were priests there to YHWH. 4. And it was the day that Elkanah sacrificed, and he gave portions to Peninnah, his wife, and to all her sons and her daughters, 5. and he gave one additional portion to Hannah because he loved Hannah and YHWH had closed her womb. 6. And, also, her rival tormented her so as to make her grieve because YHWH had closed

her womb. 7. And so it happened year after year: when she went up to the House of YHWH she would provoke her like that, and she would weep and would not eat.

8. And her husband, Elkanah, said to her, "Hannah, why are you weeping? And why don't you eat? And why is your heart in a bad way? Aren't I better to you than ten sons?"

9. And Hannah got up after eating in Shiloh and after drinking—and Eli, the priest, was sitting on a chair by the doorpost of the Temple of YHWH—10. and she was bitter-souled and prayed to YHWH and *wept*. 11. And she made a vow and said, "YHWH of hosts, if you'll look at your maid's degradation and remember me and not forget your maid and give your maid seed of men, then I'll give him to YHWH all the days of his life, and a razor won't go up on his head."

12. And it was as she kept on praying before YHWH, and Eli had been watching her mouth—13. and Hannah had been speaking in her heart: just her lips had been moving, and her voice was not being heard—and Eli thought her to be a drunk. 14. And Eli said to her, "How long will you be getting drunk? Put your wine away from you!"

15. And Hannah answered, and she said, "No, my lord, I'm a bitter-spirited woman, and I haven't drunk wine and beer. And I was spilling out my soul before YHWH. 16. Don't take your maid for a good-for-nothing, because I was speaking this long because my grief and my torment are so great."

17. And Eli answered, and he said, "Go in peace, and may the God of Israel give what you asked from Him."

18. And she said, "May your maid find favor in your eyes."

And the woman went her way. And she ate, and her face was no longer downcast. 19. And they got up early in the morning and bowed before YHWH and went back and came to their house at

Ramah. And Elkanah knew Hannah, his wife. And YHWH remembered her. 20. And it was when the time came around: and Hannah became pregnant and gave birth to a son. And she called his name Samuel "because I asked for him from YHWH."

21. And the man Elkanah and all his house went up to offer the regular sacrifice and his vow to YHWH. 22. And Hannah did not go up because, she said to her husband, "Until the boy will be weaned; then I'll bring him, and he'll appear before YHWH and live there forever."

23. And Elkanah, her husband, said to her, "Do what's good in your eyes. Stay until you've weaned him. Only let YHWH bring about His word."

And the woman stayed and nursed her son until she weaned him. 24. And she brought him up with her to Shiloh when she had weaned him, with a three-year-old bull and one ephah of flour and a bottle of wine, and she brought him to the House of YHWH at Shiloh, and the boy was young. 25. And they slaughtered the bull and brought the boy to Eli, 26. and she said, "Please, my lord, by your soul's life, my lord, I'm the woman who was standing with you here praying to YHWH. 27. I prayed for this boy, and YHWH gave me what I asked from Him. 28. And also, I've lent him to YHWH. All the days that he lives he's lent to YHWH." And he bowed there to YHWH.

2:1 And Hannah prayed and said:

My heart is happy in YHWH;
my horn is high in YHWH.
My mouth is wide over my enemies,
for I rejoice in your salvation.

2. There's no holy one like YHWH,
for there *is* no one beside you,
and there's no rock like our God.

3. Don't talk so much: high, high,
   arrogance coming from your mouth,
   for YHWH is God of knowledge,
   and deeds are gauged by Him.
4. The bows of the powerful are broken,
   and stumblers are braced with strength.
5. The full hire out for bread,
   and the hungry have stopped.
   So the childless bears seven,
   and the child-full languishes.
6. YHWH kills and gives life,
   lowers to Sheol and raises.
7. YHWH makes poor and makes rich,
   brings low and uplifts,
8. elevates poor from dust,
   lifts needy from trash,
   seating them with princes,
   bestowing on them a seat of honor.
   For the pillars of the earth are YHWH's,
   and He set the world upon them.
9. He watches over His faithful's feet,
   and the wicked are hushed in darkness,
   for not by might does a man prevail.
10. YHWH: His adversaries will be broken.
   He'll thunder at them in the skies.
   YHWH will judge the ends of the earth
   and give strength to His king
   and lift His anointed's horn.

11. And Elkanah went to Ramah, to his house, and the boy was serving YHWH before Eli the priest.

12. And Eli's sons were good-for-nothings. They did not know YHWH. 13. And the priests' rule with the people was: when any man would make a sacrifice, the priest's boy would come while the meat was cooking, with a three-pronged fork in his hand, 14. and he would strike it in the pan or kettle or cauldron or pot. The priest would take for himself everything that the fork would bring up. They did this to all of Israel who were coming there in Shiloh. 15. Also before they would burn the fat, the priest's boy would come and say to the man who was sacrificing, "Give meat to roast for the priest, since he won't take cooked meat from you, but only raw."

16. And if the man would say to him, "They'll *burn* the fat right away, and then take as your soul desires," then he would say, "No! You'll give it now. And if not I'll take it by force."

17. And the boys' sin was very big before YHWH because the people rejected the offering of YHWH. 18. And Samuel was serving before YHWH, a boy wearing a linen ephod, 19. and his mother would make him a little robe and bring it to him regularly when she would come up with her husband to make the periodic sacrifice. 20. And Eli blessed Elkanah and his wife and said, "May YHWH grant you seed from this woman in place of the one she lent to YHWH." And they went to his place. 21. And YHWH took account of Hannah, and she became pregnant and gave birth to three sons and two daughters.

And the boy Samuel grew up with YHWH.

22. And Eli was very old, and he heard all that his sons were doing to all of Israel—and that they were lying with the women

who worked at the entrance of the Tent of Meeting. 23. And he said to them, "Why do you do things like these, that I hear bad things about you from all these people? 24. Don't, my sons, because the thing that I hear is not good: causing YHWH's people to trespass. 25. If a man sins against a man, then God will appraise him; but if a man sins against YHWH, who prays for him?"

And they would not listen to their father's voice, because YHWH purposed to kill them!

26. And the boy Samuel went on growing up and being good both with YHWH and with people.

27. And a man of God came to Eli and said to him, "YHWH said this: 'Was I *revealed* to your father's house when they were in Egypt at Pharaoh's house, 28. and did I choose him as priest for me out of all Israel's tribes, to go up on my altar, to burn incense, to bear an ephod before me? And I gave your father's house all of the children of Israel's offerings made by fire. 29. Why do you kick at my sacrifice and at my offering that I commanded at my abode and honor your sons more than me, to fatten yourselves from the first of every offering of Israel for my people?' 30. Therefore, the word of YHWH, God of Israel, is: 'I had *said* your house and your father's house would walk before me forever. And now YHWH's word is: Far be it from me—because I'll honor those who honor *me,* and those who deride me will be lowered. 31. Here, days are coming when I'll cut off your arm and the arm of your father's house so that there won't be an old man in your house, 32. and you'll see a foe at my abode in everything that will benefit Israel. And there won't be an old man in your house for all time. 33. And a man of yours whom I won't cut off from my altar will be to exhaust your eyes and grieve your soul, and any proliferation of your house will die by the sword of men. 34. And this is your sign, which will happen to your two sons, to Hophni and Phinehas: the two of them

will die in one day. 35. And I'll raise a reliable priest for me, who will do what is in my heart and my soul, and I'll build a reliable house for him, and he'll walk before my anointed for all time. 36. And it will be that everyone who is left in your house will come to bow to him for a piece of silver and a loaf of bread and will say, "Attach me to one of the priesthoods so I can eat a piece of bread."'"

3:1. And the boy Samuel was serving YHWH before Eli. And YHWH's word was rare in those days; there was not widespread vision. 2. And it was on that day, and Eli was lying down in his place. And his eyes had begun to be dim; he was not able to see. 3. And the lamp of God had not yet gone out, and Samuel was lying down in the Temple of YHWH where God's ark was. 4. And YHWH called Samuel. And he said, "Here I am." 5. And he ran to Eli and said, "Here I am because you called me."

And he said, "I didn't call. Go back, lie down." And he went and lay down.

6. And YHWH went on to call Samuel again. And Samuel got up and went to Eli and said, "Here I am because you called me."

And he said, "I didn't call, my son. Go back, lie down."

7. And Samuel had not yet known YHWH, and YHWH's word had not yet been revealed to him. 8. And YHWH went on to call Samuel the third time, and he got up and went to Eli and said, "Here I am because you called me."

And Eli understood that YHWH was calling the boy. 9. And Eli said to Samuel, "Go, lie down, and it will be, if He calls you, that you will say, 'Speak, YHWH, because your servant is listening.'" And Samuel went and lay down in his place.

10. And YHWH came and stood and called as each time before, "Samuel, Samuel."

And Samuel said, "Speak, because your servant is listening."

11. And YHWH said to Samuel, "Here, I'm doing something in Israel such that the two ears of whoever hears about it will tingle. 12. On that day I'll bring about for Eli everything that I spoke about his house, beginning to end. 13. And I've told him that I'm judging his house forever for the crime that he knew that his sons were blaspheming and didn't restrain them. 14. And therefore I've sworn to Eli's house that the crime of Eli's house won't be atoned for by sacrifice and offering forever."

15. And Samuel lay down until morning, when he opened the doors of the house of YHWH. And Samuel was afraid to tell the vision to Eli. 16. And Eli called Samuel and said, "Samuel, my son."

And he said, "Here I am."

17. And he said, "What is the thing that He spoke to you? Don't hide from me. May God do this to you and double this if you hide a word from me out of every word that He spoke to you." 18. And Samuel told him every word and did not hide from him. And he said, "It is YHWH. Let him do what is good in His eyes."

19. And Samuel grew, and YHWH was with him and let none of his words fall to the ground. 20. And all Israel, from Dan to Beer-sheba, knew that Samuel was trustworthy as a prophet of YHWH. 21. And YHWH continued to appear in Shiloh, because YHWH was revealed to Samuel in Shiloh by YHWH's word, 4:1. and Samuel's word was for all Israel.

And Israel went out against the Philistines for battle; and they camped at Eben-ezer, and the Philistines camped in Aphek. 2. And the Philistines aligned against Israel. And the battle spread, and Israel was stricken before the Philistines, and they struck about four thousand men in the army in the field. 3. And the people came to the camp, and Israel's elders said, "Why did YHWH strike us before the Philistines today? Let's bring the ark of the covenant of YHWH to us from Shiloh, and let it come among us and save us from our enemies' hand." 4. And the people sent to Shiloh and carried the ark of the covenant of YHWH of Hosts Who Is Enthroned on the Cherubs. And Eli's two sons, Hophni and Phinehas, were there with the ark of God's covenant. 5. And it was when the ark of the covenant of YHWH came to the camp: and all Israel made a big shout, and the earth resounded.

6. And the Philistines heard the sound of the shout and said, "What's the sound of this big shout in the Hebrews' camp?" And they knew that YHWH's ark had come to the camp. 7. And the Philistines were afraid because, they said, "God has come to the camp." And they said, "Woe to us, because there hasn't been anything like this in previous days. 8. Woe to us! Who will rescue us from the hand of these mighty gods? These are the gods who struck Egypt with every kind of defeat in the wilderness. 9. Fortify yourselves and be men, Philistines, or else you'll serve the Hebrews as they've served you; so you'll be men and fight!" 10. And the Philistines fought, and Israel was stricken, and each man fled to his tents, and the defeat was very big, and thirty thousand foot soldiers from Israel fell. 11. And the ark of Israel was taken! And Eli's two sons, Hophni and Phinehas, died.

12. And a man of Benjamin ran from the army and came to Shiloh on that day, and his clothes were torn, and dirt was on his head. 13. And he came, and here was Eli sitting on a chair by the side of the road, watching, because his heart was trembling over God's ark. And the man came to tell it in the city, and all the city cried out. 14. And Eli heard the sound of the cry and said, "What's the sound of this commotion?" And the man hurried and came and told Eli. 15. And Eli was ninety-eight years old, and his eyes were set, and he was not able to see.

16. And the man said to Eli, "I'm the one who came from the army. And I fled from the army today."

And he said, "What was the report, my son?"

17. And the one bringing the news answered, and he said, "Israel fled before the Philistines. And also there was a big slaughter among the people. And also your two sons, Hophni and Phinehas, died. And God's ark was taken."

18a. And it was when he mentioned God's ark that he fell from the chair backward by the side of the gate, and his neck was broken, and he died, because the man was old and heavy.

19. And his daughter-in-law, Phinehas's wife, was pregnant, close to labor. And she heard the news that God's ark was taken and that her father-in-law and her husband were dead, and she bent down and gave birth because her pains came upon her. 20. And at the time of her dying those who stood by her spoke: "Don't be afraid, because you've given birth to a boy." And she did not answer, and she did not pay attention. 21. And she called the boy Ichabod, saying, "The glory has departed from Israel" on account of the taking of God's ark and on account of her father-in-law and her husband. 22. And she said, "The glory has departed from Israel because God's ark has been taken."

5:1. And the Philistines took God's ark and brought it from Eben-ezer to Ashdod. 2. And the Philistines took God's ark and brought it to the house of Dagon and set it up by Dagon. 3. And the Ashdodites got up early on the following day, and here was Dagon falling forward to the ground in front of YHWH's ark! And they took Dagon and put him back in his place. 4. And they got up early in the morning on the following day, and here was Dagon falling forward to the ground in front of YHWH's ark, and Dagon's head and his two hands were cut off, at the threshold. Only Dagon's torso was left to him. 5. On account of this, the priests of Dagon and all who come to the house of Dagon do not step on the threshold of Dagon in Ashdod to this day.

6. And YHWH's hand was heavy toward the Ashdodites, and He devastated them and struck them, Ashdod and its borders, with hemorrhoids. 7. And the people of Ashdod saw how it was, and they said, "The ark of the God of Israel will not sit among us, because His hand is hard on us and on Dagon, our god." 8. And they sent and gathered all the lords of the Philistines to them and said, "What shall we do to the ark of the God of Israel?"

And they said, "Let the ark of the God of Israel be moved to Gath." And they moved the ark of the God of Israel. 9. And it was after they moved it, and YHWH's hand was against the city—a very big terror—and He struck the people of the city, from young to old, and hemorrhoids broke out on them.

10. And they sent away God's ark to Ekron. And it was as God's ark came into Ekron, and the Ekronites cried out, saying, "They've moved the ark of the God of Israel to me, to kill me and my people!" 11. And they sent and gathered all the lords of the Philistines and said, "Send the ark of the God of Israel away, and let it go back to its place and not kill me and my people," because there was a terror of death in the whole city. God's hand was very heavy there. 12. And the people who did not die were struck with hemorrhoids, and the city's cry went up to the skies.

6:1. And YHWH's ark was in the Philistines' territory seven months. 2. And the Philistines called the priests and the diviners, saying, "What shall we do to YHWH's ark? Inform us: with what shall we send it away to its place?"

3. And they said, "If you're sending away the ark of the God of Israel, don't send it away empty, but *pay back* a guilt offering to Him. Then you'll be healed, and it will be known to you why His hand doesn't turn away from you."

4. And they said, "What is the guilt offering that we shall pay Him back?"

And they said, "The number of the lords of the Philistines: five gold hemorrhoids and five gold mice, because it's one plague on everyone and on your lords. 5. And you shall make images of your hemorrhoids and images of your mice that are destroying the land, and give glory to Israel's God. Maybe He'll lighten His hand from on you and from on your gods and from on your land. 6. And why do you make your heart heavy as Egypt and Pharaoh made their heart heavy? When He abused them, didn't they let them go, and they went? 7. And now, take and prepare one new wagon and two milking cows on which there's never been a yoke, and hitch the cows to the wagon and take their calves behind them back home. 8. And take YHWH's ark and place it into the wagon and set the gold objects that you're paying back to Him as a guilt offering in a chest by its side and send it back, and let it go. 9. And you'll see: if it goes up the way to its own border, to Beth-shemesh, He did this big bad thing to us; and if not, then we'll know that His hand didn't plague us, it was a chance occurrence that we had."

10. And the people did this. They took two milking cows and hitched them to the wagon and shut up their calves at home, 11. and they set YHWH's ark into the wagon, and the chest and the gold mice and the images of their hemorrhoids. 12. And the cows went straight down the road: on the road to Beth-shemesh. They went on one highway, lowing as they went, and did not turn right or left. And the lords of the Philistines were going after them to the border of Beth-shemesh. 13. And Beth-shemesh was reaping the wheat harvest in the valley, and they raised their eyes and saw the ark and were happy to see it. 14. And the wagon came to the field of Joshua, a Beth-shemeshite, and stood there. And a big stone was there, and they broke up the wood of the wagon and offered the

cows as a burnt offering to YHWH. 15. And the Levites brought down YHWH's ark and the chest that was with it that had the gold objects in it, and they set it on the big rock, and the people of Beth-shemesh made burnt offerings and sacrifices on that day to YHWH. 16. And the five lords of the Philistines saw and went back to Ekron on that day.

17. And these are the gold hemorrhoids that the Philistines paid back as a guilt offering to YHWH: one for Ashdod, one for Gaza, one for Ashkelon, one for Gath, one for Ekron. 18. And the gold mice: the number of all the cities of the Philistines belonging to the five lords, from the fortified city to the unwalled village, and to the big rock on which they had rested YHWH's ark, to this day in the field of Joshua, the Beth-shemeshite.

19. And He struck among the people of Beth-shemesh because they looked in YHWH's ark. And He struck fifty thousand seventy men of the people, and the people mourned because YHWH had struck a big blow among the people. 20. And the people of Beth-shemesh said, "Who is able to stand before YHWH, this holy God, and for whom shall He go up from upon us?" 21. And they sent messengers to those who lived in Kiriath-jearim, saying, "The Philistines have given back YHWH's ark. Come down. Bring it up to you." 7:1. And the people of Kiriath-jearim came and brought up YHWH's ark and brought it to Abinadab's house on the hill, and they sanctified Eleazar, his son, to watch over YHWH's ark. 2. And it was, from the day the ark stayed in Kiriath-jearim, that many days passed, and they were twenty years, and all the house of Israel felt grief toward YHWH.

5. And Samuel said, "Gather all Israel to Mizpeh, and I'll pray for you to YHWH."

6. And they gathered to Mizpeh and drew water and spilled it out before YHWH and fasted on that day and said there, "We've

sinned against YHWH." And Samuel judged the children of Israel in Mizpeh. 7. And the Philistines heard that the children of Israel had gathered at Mizpeh, and the lords of the Philistines went up against Israel. And the children of Israel heard and were afraid of the Philistines. 8. And the children of Israel said to Samuel, "Don't keep quiet from us, from crying out to YHWH, our God, that He'll save us from the Philistines' hand." 9. And Samuel took one lamb suckling and offered it whole as a burnt offering to YHWH, and Samuel cried out to YHWH for Israel. And YHWH answered him. 10. And Samuel was making the burnt offering, and the Philistines approached for war against Israel. And YHWH thundered with a big sound in that day on the Philistines and threw them into tumult, and they were stricken before Israel. 11. And the people of Israel went out from Mizpeh and pursued the Philistines and struck them up to the area below Beth-car.

12. And Samuel took one rock and set it between Mizpeh and Shen and called its name Eben-ezer and said, "YHWH helped us up to here." 13. And the Philistines were subdued and did not continue to come in Israel's border anymore. And YHWH's hand was against the Philistines all of Samuel's days. 14. And the cities that the Philistines had taken from Israel went back to Israel, from Ekron to Gath, and Israel rescued their borders from the Philistines' hand. And there was peace between Israel and the Amorite. 15. And Samuel judged Israel all the days of his life, 16. and he went from year to year and circuited Beth-El and Gilgal and Mizpeh and judged Israel in all these places. 17. And his place of return was Ramah because his house was there, and he judged Israel there and built an altar to YHWH there.

8:1. And it was when Samuel was old: and he set his sons as judges for Israel. 2. And his firstborn son's name was Joel, and his second's name was Abijah, judging in Beer-sheba. 3a. And his sons did not go in his ways, and they reached out after profit. 4. And all of Israel's elders gathered and came to Samuel at Ramah, 5. and said to him, "Here *you* have become old, and your sons haven't gone in your ways. Now set us a king to judge us like all the nations." 6. And the thing was bad in Samuel's eyes, when they said, "Give us a king to judge us." And Samuel prayed to YHWH.

7. And YHWH said to Samuel, "Listen to the people's voice, to everything that they say to you, because it's not you whom they've rejected, but they've rejected *me* from ruling over them. 9b. But you shall tell them the law of the king who will rule over them."

10. And Samuel said all of YHWH's words to the people who were asking for a king from him. 11. And he said, "This will be the law of the king who will rule over you: He'll take your sons and set them for himself, for his chariot and for his horsemen, and they'll run in front of his chariot. 12. And he'll set officers of thousands and officers of fifties for himself, and to do his plowing and to reap his harvest and to make his weapons of war and his chariot equipment. 13. And he'll take your daughters, for perfumers and for cooks and for bakers. 14. And he'll take your best fields and vineyards and olives, and he'll give them to his servants. 15. And he'll take a tenth of your seed and your vineyards and give them to his officers and to his servants. 16. And he'll take your servants and your maids and your choice young men, your best ones, and your asses and use them for his work. 17. He'll take a tenth of your sheep. And *you* will become slaves to him. 18. And you'll cry out in that day because of your king, whom you chose for yourselves, and YHWH won't answer you in that day."

19. And the people refused to listen to Samuel's voice, and they said, "No, but a king will be over us, 20. and we, too, will be like all

the nations, and our king will judge us and go out in front of us and fight our battles."

21. And Samuel listened to all the people's words and spoke them in YHWH's ears. 22a. And YHWH said to Samuel, "Listen to their voice and give them a king."

10:17. And Samuel called the people together to YHWH at Mizpeh, 18. and said to the children of Israel, "YHWH, God of Israel, has said this: 'I brought up Israel from Egypt and rescued you from Egypt's hand and from the hand of all the kingdoms that were oppressing you.' 19. And today you've rejected your God, who is a savior for you from all your bad situations and your troubles, and you said to him, 'You should set a king over us.' And now, stand before YHWH by your tribes and by your thousands." 20. And Samuel brought all of Israel's tribes forward, and the tribe of Benjamin was taken by lot. 21. And he brought the tribe of Benjamin forward by its families, and the family of Matri was taken by lot. And Saul, son of Kish, was taken.

And they looked for him, and he was not found. 22. And they asked further of YHWH, "Has the man come here?"

And YHWH said, "Here he is, hidden among the equipment." 23. And they ran and took him from there, and he stood up among the people, and he was taller than all the people from his shoulder and up.

24. And Samuel said to all the people, "Have you seen the one whom YHWH has chosen, that there's none like him among all the people?!"

And all the people shouted and said, "Long live the king!"

25. And Samuel spoke the law of the monarchy to the people and wrote it in a scroll and laid it before YHWH. And Samuel sent all the people away, each to his house. 26. And Saul, too, went to his house at Gibeah, and the warriors whose heart God had touched

went with him. 27. And good-for-nothings said, "What, will this one save us?!" And they disdained him and did not bring him any gift. And he kept quiet.

11:1. And Nahash the Ammonite went up and camped against Jabesh-Gilead. And all the people of Jabesh said to Nahash, "Make a covenant with us, and we'll serve you."

2. And Nahash the Ammonite said to them, "I'll make it with you on this condition: the gouging of every right eye of yours, and I'll make it a disgrace on all of Israel."

3. And Jabesh's elders said to him, "Hold off from us for seven days, and let us send messengers in every border of Israel; and if no one saves us then we'll come out to you." 4. And the messengers came to Gibeah of Saul and spoke the words in the people's ears, and all the people raised their voices and wept.

5. And here was Saul coming behind the oxen from the field, and Saul said, "What's with the people that they're weeping?" And they told him the words of the people of Jabesh. 6. And God's spirit rushed onto Saul when he heard these things, and his anger flared very much, 7. and he took a yoke of oxen and cut it up and sent it in every border of Israel by the hand of the messengers, saying, "Whoever doesn't come out behind Saul and behind Samuel: this will be done to his oxen." And the fear of YHWH fell upon the people, and they came out like one man. 8. And he mustered them in Bezek, and the children of Israel were three hundred thousand, and the men of Judah thirty thousand.

9. And they said to the messengers who came, "Say this to the men of Jabesh-Gilead: 'You'll have salvation tomorrow when the sun is hot.'" And the messengers came and told the people of Jabesh-Gilead, and they were happy.

10. And the people of Jabesh said, "We'll come out to you tomorrow, and you'll do to us whatever is good in your eyes."

11. And it was on the following day, and Saul set the people in three companies, and they came into the camp in the morning watch and struck Ammon until the heat of the day, and those who were left were scattered, and no two of them were left together.

12. And the people said to Samuel, "Who was it who said, 'Shall Saul rule us?!' Give us the people, and we'll kill them!"

13. And Saul said, "Not a man will be killed on this day, because today YHWH made salvation in Israel." 14. And Samuel said to the people, "Come on, and let's go to Gilgal and renew the monarchy there." 15. And all the people went to Gilgal and made Saul king there before YHWH in Gilgal and made peace-offering sacrifices there before YHWH. And Saul and all the people of Israel were very happy there.

12:1. And Samuel said to all Israel, "Here, I've listened to your voice, to everything that you said to me, and I've made a king over you. 2. And now, here is the king walking before you. And I've grown old and gray, and here are my sons with you, and I've walked before you from my youth to this day. 3. Here I am. Respond against me before YHWH and before His anointed: whose ox have I taken, and whose ass have I taken, and whom have I defrauded? Whom have I oppressed, and from whose hand have I taken a bribe and hidden my eyes due to it—and I'll pay it back to you."

4. And they said, "You haven't defrauded us and haven't oppressed us and haven't taken anything from any man's hand."

5. And he said to them, "YHWH is a witness against you, and His anointed is a witness this day, that you didn't find anything in my hand."

And they said, "Witness."

6. And Samuel said to the people, "YHWH who made Moses and Aaron and who brought your fathers up from the land of Egypt is witness. 16. Now stand still and see this big thing that YHWH is doing before your eyes. 17. Isn't it wheat harvest today? I'll call to YHWH, and He'll give thunderings and rain. And know and see that your bad thing that you've done—to ask for a king for yourselves—is tremendous in YHWH's eyes." 18. And Samuel called to YHWH, and YHWH gave thunderings and rain on that day. And all the people feared YHWH and Samuel very much.

19. And all the people said to Samuel, "Pray for your servants to YHWH, your God, and let us not die, because we've added bad to all our sins to ask for a king for ourselves."

20a. And Samuel said to the people, 22. "YHWH won't forsake His people for His great name's sake, because YHWH was pleased to make you His people. 23. I, too: far be it from me to sin against YHWH, to stop praying for you. And I'll teach you the good and right way."

15:1. And Samuel said to Saul, "YHWH sent me to anoint you as king over His people, over Israel, so now listen to the voice, YHWH's words. 2. YHWH of hosts said this: 'I have noted what Amalek did to Israel, that he set against him in the way when he came up from Egypt. 3. Now go and strike Amalek and completely destroy everything that he has and don't spare it, and kill them, from man to woman, from child to suckling, from ox to sheep, from camel to ass.'"

4. And Saul assembled the people and mustered them in Telaim, two hundred thousand foot soldiers, and ten thousand men of Judah. 5. And Saul came up to the city of Amalek and lay in wait in the wadi. 6. And Saul said to the Kenites, "Go, turn away, go down from among the Amalekites, in case I'd destroy you with him, for you showed kindness to all the children of Israel when

they came up from Egypt." And the Kenites turned away from among Amalek. 7. And Saul struck Amalek from Havilah as you come to Shur, which is facing Egypt. 8. And he captured Agag, king of Amalek, alive, and he completely destroyed the whole people by the sword. 9. And Saul and the people spared Agag and the best of the sheep and the oxen and the fatlings and the lambs and everything that was good, and they were not willing to destroy them completely; but they completely destroyed everything that was despised and spoiled.

10. And YHWH's word came to Samuel, saying, 11. "I regret that I made Saul king, because he has gone back from following me and hasn't carried out my words." And Samuel was angry, and he cried out to YHWH all night. 12. And Samuel got up early to meet Saul in the morning.

And it was told to Samuel, saying, "Saul has come to Carmel, and, here, he erected a monument for himself and went back and crossed over and went down to Gilgal."

13. And Samuel came to Saul, and Saul said to him, "You are blessed to YHWH. I've carried out YHWH's words."

14. And Samuel said, "And what is this sound of sheep in my ears and this sound of oxen that I hear?"

15. And Saul said, "They've brought them from the Amalekite, because the people spared the best of the sheep and the oxen in order to sacrifice to YHWH, your God, and we completely destroyed the rest."

16. And Samuel said to Saul, "Hold off, and let me tell you what YHWH spoke to me this night."

And he said to him, "Speak."

17. And Samuel said, "Is it not, if you're little in your own eyes, you're head of the tribes of Israel? And YHWH anointed you as king over Israel. 18. And YHWH sent you on a path and said, 'Go

and completely destroy those who have sinned, the Amalekite, and battle against him until you've finished them.' 19. And why didn't you listen to YHWH's voice, that you dashed to the spoil and did what is bad in YHWH's eyes?"

20. And Saul said to Samuel, "But I did listen to YHWH's voice, and I went on the path on which YHWH sent me, and I brought Agag, king of Amalek, and I completely destroyed Amalek. 21. And the people took sheep and oxen from the spoil, the first of what was supposed to be completely destroyed, to sacrifice to YHWH, your God, in Gilgal."

22. And Samuel said, "Does YHWH have pleasure in burnt offerings and sacrifices as much as *listening* to YHWH's voice? Here:

> listening is better than sacrifice,
> paying attention than rams' fat;
> 23. rebellion is like the sin of divination,
> and arrogance: iniquity and teraphim;
> because you rejected YHWH's word,
> He has rejected you from being king."

24. And Saul said to Samuel, "I've sinned, because I violated YHWH's mouth and your words because I was afraid of the people, and I listened to their voice. 25. And now, bear my sin and come back with me, and I'll bow to YHWH."

26. And Samuel said to Saul, "I won't go back with you, because you rejected YHWH's word, and YHWH has rejected you from being king over Israel." 27. And Samuel turned to go, and Saul held onto the corner of his robe, and it was ripped. 28. And Samuel said to him, "YHWH has ripped the kingdom of Israel away from you today and will give it to your companion who's better than you.

29. And, also, Israel's Everlasting One won't lie and won't regret, because He's not a human that He should regret."

30. And he said, "I've sinned. Now honor me in front of my people's elders and in front of Israel and come back with me, and I'll bow to YHWH, your God." 31. And Samuel went back after Saul, and Saul bowed to YHWH.

32. And Samuel said, "Bring Agag, king of Amalek, to me."

And Agag went to him with pleasure, and Agag said, "The bitterness of death has surely gone by."

33. And Samuel said, "As your sword has bereaved women, so your mother will be bereaved among women." And Samuel executed Agag before YHWH in Gilgal.

34. And Samuel went to Ramah, and Saul went up to his house at Gibeah of Saul. 35. And Samuel did not see Saul again to the day of his death, because Samuel mourned for Saul. And YHWH regretted that He had made Saul king over Israel. 16:1. And YHWH said to Samuel, "How long will you be mourning for Saul? And I've rejected him from ruling Israel. Fill your horn with oil and go. I'll send you to Jesse, the Bethlehemite, because I've seen a king for me among his sons."

2. And Samuel said, "How shall I go? And Saul will hear, and he'll kill me!"

And YHWH said, "Take a heifer of the herd in your hand and say, 'I've come to sacrifice to YHWH,' 3. and call Jesse to the sacrifice. And I'll make known to you what you'll do, and you'll anoint for me the one I say to you." 4. And Samuel did what YHWH spoke, and he came to Bethlehem.

And the city's elders trembled to meet him and said, "You come in peace?"

5. And he said, "Peace. I've come to sacrifice to YHWH. Consecrate yourselves and come with me to sacrifice. And he consecrated

Jesse and his sons and called them to the sacrifice. 6. And it was when they were coming: and he saw Eliab, and he said, "YHWH's anointed is just in front of Him!"

7. And YHWH said to Samuel, "Don't look at his appearance or at how tall he stands, because I've rejected him; because it's not what the human sees, because the human sees what's in front of the eyes, and YHWH sees what's in the heart."

8. And Jesse called Abinadab and had him pass in front of Samuel, and he said, "YHWH hasn't chosen this one either." 9. And Jesse had Shammah pass, and he said, "YHWH hasn't chosen this one either." 10. And Jesse had his seven sons pass in front of Samuel, and Samuel said to Jesse, "YHWH hasn't chosen these." 11. And Samuel said to Jesse, "Was that the end of the boys?"

And he said, "The youngest is still left; and, here, he's tending the sheep."

And Samuel said to Jesse, "Send and get him, because we won't turn away until he comes here." 12. And he sent and brought him. And he was all ruddy, with beautiful eyes, and good-looking.

And YHWH said, "Get up, anoint him, because this is the one." 13. And Samuel took the horn of oil and anointed him among his brothers. And YHWH's spirit rushed to David from that day on. And Samuel got up and went to Ramah.

14. And YHWH's spirit turned away from Saul, and a bad spirit from YHWH terrified him. 15. And Saul's servants said to him, "Here, a bad spirit of God is terrifying you. 16. Let our lord say that your servants before you should seek a man who knows how to play the lyre, and it will be that when the bad spirit of God is on you he'll play it in his hand, and it will be good for you."

17. And Saul said to his servants, "Look for a man for me who plays well and bring him to me."

18. And one of the young men answered, and he said, "Here, I've seen a son of Jesse, the Bethlehemite, who knows how to play and is a powerful warrior and a man of war, well spoken and a handsome man, and YHWH is with him."

19. And Saul sent messengers to Jesse, and he said, "Send me David, your son, who's with the sheep." 20. And Jesse took an ass-load of bread and a wineskin and one goat kid and sent them by the hand of David, his son, to Saul. 21. And David came to Saul and stood before him, and he liked him very much, and he became his armor-bearer. 22. And Saul sent to Jesse, saying, "Let David stand before me, because he's found favor in my eyes." 23. And it was, when a spirit of God was on Saul, that David would take the lyre and play it in his hand, and Saul had relief and felt well, and the bad spirit turned away from him.

17:1. And the Philistines gathered their camps for war. And they were gathered at Sochoh, which was Judah's, and camped between Sochoh and Azekah at Ephes-dammim. 2. And Saul and the men of Israel were gathered and camped in the valley of Elah and aligned for war against the Philistines. 3. And the Philistines were standing on the mountain on one side, and Israel were standing on the mountain on the other side, and the valley was between them. 4. And a man of the land between, from the Philistine camps, came out. Goliath was his name, from Gath. His height was six cubits and a span. 5. And he had a bronze helmet on his head and was wearing a coat of mail, and the weight of the coat was five thousand shekels of bronze, 6. and bronze greaves on his legs and a bronze javelin between his shoulders, and his spear's shaft was like a weavers' beam, 7. and his spear's blade was six hundred shekels of iron, and a shield-bearer was going in front of him. 8. And he stood and called to Israel's lines and said to them, "Why do you come out to align for war? Am I not the Philistine and you Saul's servants? Choose yourselves a man, and let him come down to me. 9. If he'll be able to fight with me and strike me down, then we'll be your servants, and if I prevail against him and strike him down, then you'll be our servants and serve us." 10. And the Philistine said, "I've disgraced Israel's lines this day! Give me a man, and let's fight together!" 11. And Saul and all Israel heard these words of the Philistine, and they were dismayed and were very afraid. 50. But David overpowered the Philistine with a sling and stone, and he struck the Philistine and killed him. And there was not a sword in David's hand!

18:6. And it was when they were coming, when David was coming back from striking the Philistine: and the women came out from all of Israel's cities for singing and dances toward Saul, the king, with drums, joyfully, and with triangles. 7. And the women who were performing sang, and they said:

> Saul has struck his thousands
> and David his ten thousands.

8. And Saul was very angry, and this thing was bad in his eyes, and he said, "They gave David the ten thousands, and to *me* they gave the thousands! And he'll yet have the kingdom as well!" 9. And Saul was eyeing David from that day on.

10. And it was on the following day, and a bad spirit of God rushed to Saul, and he was frenzied inside the house. And David was playing it in his hand as on other days. And a spear was in Saul's hand. 11. And Saul hurled the spear and said, "I'll strike through David and through the wall!" And David spun away in front of him twice. 12. And Saul was afraid of David because

YHWH was with him and had turned from Saul. 13. And Saul removed him from being with him. And he made him an officer over a thousand, and he went out and came in before the people. 16. And all of Israel and Judah loved David because he was going out and coming in before them.

20. And Michal, Saul's daughter, loved David. And they told Saul, and the thing was all right in his eyes! 21a. And Saul said, "Let me give her to him so that she'll be a snare for him, so the Philistines' hand will be against him." 22. And Saul commanded his servants, "Speak to David secretly, saying, 'Here, the king is pleased with you, and all his servants love you, and now become the king's son-in-law.'"

23. And Saul's servants spoke these things in David's ears. And David said, "Is becoming the king's son-in-law an inconsequential thing in your eyes, when I'm a poor and inconsequential man?"

24. And Saul's servants told him, saying, "David spoke things like these."

25. And Saul said, "Say this to David, 'The king has no pleasure in a bride price but rather in a hundred foreskins of Philistines to be avenged against the king's enemies.'" And Saul was thinking of making David fall by the Philistines' hand. 26. And his servants told these things to David, and the thing was right in David's eyes in order to become the king's son-in-law; and not many days were amassed, 27. and David got up and went, he and his men, and struck two hundred men of the Philistines. And David brought their foreskins and amassed them for the king to become the king's son-in-law. And Saul gave him Michal, his daughter, as a wife. 28. And Saul saw and knew that YHWH was with David, and Michal, Saul's daughter, loved him, 29. and Saul proceeded to be more afraid of David. And Saul was an enemy to David for all time.

19:1. And Saul spoke to Jonathan, his son, and to all his servants to kill David. But Jonathan, son of Saul, was very pleased with David, 2. and Jonathan told David, saying, "Saul, my father, is seeking to kill you. And now, watch yourself in the morning, and sit in a concealed place and hide yourself. 3. And I'll come out, and I'll stand by my father in the field where you'll be, and I'll speak about you to my father, and I'll see what's what, and I'll tell you."

4. And Jonathan spoke well about David to Saul, his father, and said to him, "Let the king not sin against his servant, against David, because he hasn't sinned against you, and because his actions have been very good for you, 5. and he took his life in his hand and struck down the Philistine, and YHWH made a big salvation for all Israel. You saw and were happy. And why should you sin against innocent blood, to kill David for nothing?"

6. And Saul listened to Jonathan's voice, and Saul swore, "As YHWH lives, he won't be killed."

7. And Jonathan called David, and Jonathan told him all these things, and Jonathan brought David to Saul, and he was before him as in previous days.

8. And there continued to be war. And David went out and fought against the Philistines and struck a big blow against them, and they fled from him.

9. And a bad spirit of YHWH was on Saul, and he was sitting in his house. And his spear was in his hand. And David was playing it in his hand. 10. And Saul sought to strike through David and through the wall with his spear. And he got away from Saul, and he struck through the wall with the spear, and David fled and escaped that night. 11. And Saul sent messengers to David's house to watch him and to kill him in the morning. And Michal, his wife, told David, saying, "If you don't save your life tonight you'll be killed tomorrow." 12. And Michal let David down through the window,

and he went and fled and escaped. 13. And Michal took the teraphim and put them in the bed and put a goat-hair pillow at its head and covered it with clothing. 14. And Saul sent messengers to take David, and she said, "He's sick."

15. And Saul sent the messengers to see David, saying, "Bring him up to me in the bed in order to kill him." 16. And the messengers came, and here were the teraphim in the bed, and the goats-hair pillow at its head. 17. And Saul said to Michal, "Why did you deceive me like this and let my enemy go so he escaped?"

And Michal said to Saul, "He said to me, 'Let me go; why should I kill you?'"

18. And David fled and escaped and came to Samuel at Ramah and told him everything that Saul had done to him. And he went, he and Samuel, and they lived in Naioth. 19. And it was told to Saul, saying, "Here, David's at Naioth in Ramah." 20. And Saul sent messengers to take David, and they saw a band of prophets prophesying and Samuel standing, stationed over them. And a spirit of God was on Saul's messengers, and they prophesied, too. 21. And they told Saul, and he sent other messengers, and they prophesied, too. And Saul proceeded to send a third set of messengers, and they prophesied, too. 22. And he, too, went to Ramah. And he came to the big cistern that was in Secu and asked and said, "Where are Samuel and David?"

And one said, "Here, at Naioth in Ramah."

23. And he went there, to Naioth in Ramah. And a spirit of God was on him, too, and he kept on going and prophesied until he came into Naioth in Ramah. 24. And he, too, stripped off his clothes; and he, too, prophesied in front of Samuel, and he fell down naked all that day and all night. On account of this they say, "Is Saul, too, among the prophets?!"

25:1. And Samuel died. And all Israel gathered and mourned him and buried him at his house in Ramah. And David got up and went down to the wilderness of Paran. 2. And there was a man in Maon, and his trade was in Carmel, and the man was very big, and he had three thousand sheep and a thousand goats. And it was when he was shearing his sheep in Carmel. 3. And the man's name was Nabal, and his wife's name was Abigail. And the woman had good sense and an attractive figure, and the man was hard and bad in his practices. And he was a Calebite. 4. And David heard in the wilderness that Nabal was shearing his sheep, 5. and David sent ten young men, and David said to the young men, "Go up to Carmel and come to Nabal and ask him in my name how he is 6. and say this: 'To life, and peace to you, and peace to your house, and peace to all that you have. 7. And now I've heard that they're shearing for you. Now, your shepherds were with us. We didn't injure them, and nothing of theirs was missing all the days they were in Carmel. 8. Ask your young men, and they'll tell you. And let the young men find favor in your eyes, because we've come on a good day. Give whatever your hand chooses to your servants and to your son, David.'"

9. And David's young men came and spoke according to all these words to Nabal in David's name. And they paused. 10. And Nabal answered David's servants and said, "Who is David, and who is son-of-Jesse? These days there are a lot of servants breaking away, each one from his lord! 11. And shall I take my bread and my water and my meat that I've slaughtered for my shearers and give it to people about whom I don't know where they're from?!" 12. And David's young men turned back to their way, and they went back and came and told him all these things.

13. And David said to his men, "Every man put on his sword!" And every man put on his sword. And David put on his sword, too. And about four hundred men went up after David, and two hundred stayed with the equipment.

14. And one of the young men told Abigail, Nabal's wife, saying, "Here, David sent messengers from the wilderness blessing our lord, and he shrieked at them! 15. And the people were very good to us, and we weren't injured, and we weren't missing anything all the days that we went around with them when we were in the field. 16. They were a wall for us, both night and day, all the days we were with them tending the sheep. 17. And now, know and see what you'll do, because a bad thing has worked out toward our lord and on his whole house, and he's too much of a good-for-nothing for anyone to speak to him!"

18. And Abigail hurried and took two hundred loaves of bread and two bottles of wine and five prepared sheep and five measures of parched grain and a hundred raisin clusters and two hundred fig cakes, and she put them on the asses. 19. And she said to her young men, "Pass ahead of me. Here I am, coming behind you." And she did not tell her husband, Nabal. 20. And it was: she was riding on the ass and going down in a concealed place of the mountain, and here were David and his men coming down toward her, and she met them.

21. And David was saying, "It was just for a betrayal that I watched over all that this one had in the wilderness—and not a thing of all that he had was missing—and he paid me back bad for good. 22. May God do this to David and double this if, by the morning light, I leave anyone who pisses against the wall out of all he has."

23. And Abigail saw David, and she hurried and got down from the ass and fell on her face and bowed to the ground in front of

David. 24. And she fell at his feet and said, "My lord, the crime is in me, myself. And let your maid speak in your ears, and listen to your maid's words. 25. Let my lord not pay attention to this good-for-nothing man, Nabal; because as his name is, that's how he is! His name is Nabal, and foolishness is with him. And I, your maid, didn't see my lord's young men whom you sent. 26. And now, my lord, as YHWH lives and as your soul lives—since YHWH has held you back from coming into blood and from avenging yourself by your hand—and now, may your enemies and those who seek bad things toward my lord be like Nabal! 27. And now, this blessing that your maid has brought to my lord: and it will be given to the young men who follow at my lord's feet. 28. Bear your maid's offense, because YHWH *will make* a trustworthy house for my lord because my lord fights YHWH's battles, and bad hasn't been found in you in all your days. 29. And a human has risen to pursue you and to seek your life, but my lord's soul will be bound in the binding of life with YHWH, your God, and He'll sling your enemies' soul in the pouch of a sling. 30. And it will be when YHWH will do for my lord everything that He has spoken, a good thing upon you, and will command that you be the designated king over Israel, 31. that this won't be a tottering for you and a stumbling of heart for my lord, to have spilled blood for nothing and for my lord to have saved himself, and YHWH will be good to my lord, and you'll remember your maid."

32. And David said to Abigail, "Blessed is YHWH, God of Israel, who sent you to me this day, 33. and blessed is your judgment, and blessed are you for keeping me from coming into blood and avenging myself by my hand. 34. And indeed, as YHWH, God of Israel, lives, who has held me back from doing you any bad, if you hadn't hurried and come to me, Nabal wouldn't have had

anyone who pisses against the wall left by morning light." 35. And David took what she had brought him from her hand and said to her, "Go up to your house in peace. See, I've listened to your voice and respected your presence."

36. And Abigail came to Nabal, and here he was, having a feast in his house like a king's feast, and Nabal's heart was in a good mood, and he was very drunk, and she did not tell him a word, small or big, until the morning light. 37. And it was in the morning, when the wine had gone out of Nabal, and his wife told him these things. And his heart died inside him, and he turned to stone. 38. And it was about ten days, and YHWH struck Nabal, and he died.

39. And David heard that Nabal was dead, and he said, "Blessed is YHWH, who has vindicated the case of my disgrace at the hand of Nabal, and has held back his servant from doing bad, and has paid back Nabal's bad on his head!"

And David sent and spoke to Abigail of taking her as his wife. 40. And David's servants came to Abigail at Carmel and spoke to her saying, "David sent us to you to take you as his wife."

41. And she got up and bowed, nose to the ground, and said, "Here your maid is as a servant to wash my lord's servants' feet." 42. And Abigail hurried and got up and rode on an ass, with her five young women who were going after her, and she went after David's messengers and became his wife.

43. And David took Ahinoam of Jezreel. And the two of them became his wives as well. 44. And Saul had given Michal, his daughter, David's wife, to Palti, son of Laish, who was from Gallim!

26:1. And the Ziphites came to Saul at Gibeah, saying, "Isn't David hiding at the hill of Hachilah, facing Jeshimon?" 2. And Saul got up, and he went down to the wilderness of Ziph, and three thousand chosen men of Israel with him, to seek David in the wilderness of Ziph. 3. And Saul camped in the hill of Hachilah, which is facing Jeshimon on the road, but David was living in the wilderness, and he saw that Saul was coming to the wilderness after him, 4. and David sent spies and knew for certain that Saul had come. 5. And David got up and came to the place where Saul had camped. And David saw the place where Saul and Abner, son of Ner, the commander of his army, had lain down, with Saul lying in the trench and the people camping around him.

6. And David answered, and he said to Ahimelech, the Hittite, and to Abishai, son of Zeruiah, brother of Joab, saying, "Who will go down with me to Saul, to the camp?"

And Abishai said, "I'll go down with you."

7. And David and Abishai came to the people by night, and here was Saul lying asleep in the trench with his spear stuck in the ground by his head and Abner and the people lying around him. 8. And Abishai said to David, "God has shut your enemy up in your hand today. And now, let me strike him with the spear and into the ground: one time, and I won't need to repeat it."

9. And David said to Abishai, "Don't destroy him, because who has put out his hand against YHWH's anointed and been held innocent?!" 10. And David said, "As YHWH lives, YHWH will strike him, or his time will come and he'll die, or he'll go down into battle and be annihilated. 11. Far be it from me, by YHWH, to put out my hand against YHWH's anointed. And now, take the spear that's by his head and the jug of water, and let's go." 12. And David took the spear and the jug of water by Saul's head, and they went. And there was no one who saw and no one who knew and

no one who woke up, because all of them were sleeping, because a slumber from YHWH had descended on them. 13. And David passed over to the other side and stood on the top of the mountain far off; the distance between them was great. 14. And David called to the people and to Abner, son of Ner, saying, "Won't you answer, Abner?"

And Abner answered, and he said, "Who are you that you've called to the king?"

15. And David said to Abner, "Are you not a man? And who is like you in Israel? So why didn't you watch over your lord, the king? Because one of the people came to destroy the king, your lord! 16. This thing that you've done is not good. As YHWH lives, you are people who deserve death because you didn't watch over your lord, over YHWH's anointed. And now, look: where are the king's spear and the jug of water that were by his head?!"

17. And Saul recognized David's voice and said, "Is that your voice, my son, David?"

And David said, "My voice, my lord the king." 18. And he said, "Why is this that my lord pursues his servant? For what have I done, and what bad thing is in my hand? 19. And now, let my lord the king hear his servant's words: If YHWH has incited you against me, let Him smell an offering; but if humans, they are cursed before YHWH because they drove me out today from joining in YHWH's legacy, saying, 'Go serve other gods.' 20. And now, let my blood not fall to the ground before YHWH's face, because the king of Israel has gone out to seek one flea, the way one pursues a partridge in the mountains."

21. And Saul said, "I've sinned. Come back, my son, David, because I won't wrong you anymore, on account of the fact that my life was precious in your eyes on this day. Here, I've been foolish and very much misguided."

22. And David answered, and he said, "Here's the king's spear, and let one of the young men pass over and get it. 23. And YHWH will pay back each man his virtue and his faithfulness, because YHWH delivered you in my hand today, and I wasn't willing to put out my hand against YHWH's anointed. 24. And, here, as your life was important in my eyes this day, so let my life be important in YHWH's eyes, and let Him rescue me from all trouble."

25. And Saul said to David, "You're blessed, my son, David. You'll both *do* things and *prevail.*"

And David went his way, and Saul went back to his place. 27:1. And David said in his heart, "Now I'll be annihilated in Saul's hand one day. I have nothing better than that I should *escape* to the land of the Philistines, and Saul will desist from me, to seek me anymore in all of Israel's border, and I'll escape from his hand." 2. And David got up and passed over, he and six hundred men who were with him, to Achish, son of Maoch, king of Gath. 3. And David lived with Achish in Gath, he and his people, each man and his household, David and his two wives, Ahinoam the Jezreelite and Abigail, Nabal's wife, the Carmelite. 4. And it was told to Saul that David had fled to Gath, and he did not go on to seek him again.

5. And David said to Achish, "If I've found favor in your eyes, let them give me a place in one of the cities of the country, and let me live there, because why should your servant live in the royal city with you?" 6. And Achish gave him Ziklag on that day. Therefore Ziklag came to belong to the kings of Judah to this day. 7. And the number of days that David lived in the Philistines' country was a year and four months. 8. And David and his men went up and raided the Geshurites and the Gezrites and the Amalekites, because they were the residents of the land from olden times as you come to Shur and as far as the land of Egypt. 9. And David struck

the land and did not leave a man or woman alive, and he took sheep and cattle and asses and camels and clothing, and he went back and came to Achish.

10. And Achish said, "Where did you raid today?"

And David said, "At the Negeb of Judah and at the Negeb of the Jerahmeelite and at the Negeb of the Kenite." 11. And David had not left a man or woman alive to bring to Gath, saying, "In case they would tell on us." This is what David did, and this was his rule all the days that he lived in the Philistines' country.

12. And Achish trusted David, saying, "He's made himself *odious* among his people in Israel and has become mine as a servant forever."

28:1. And it was in those days, and the Philistines gathered their army camps for war against Israel. And Achish said to David, "You *know* that you'll go out with me in the camp, you and your men."

2. And David said to Achish, "In that case, you know what your servant will do."

And Achish said to David, "In that case, I'll make you watchman over my head for all time."

3. And Samuel had died, and all Israel mourned him, and they buried him in Ramah, in his city. And Saul had turned the mediums and necromancers out of the land. 4. And the Philistines were gathered and came and camped in Shunem; and Saul gathered all of Israel, and they camped at Gilboa. 5. And Saul saw the Philistines' camp and was afraid, and his heart trembled very much. 6. And Saul asked of YHWH, and YHWH did not answer him, neither through dreams nor through Urim nor through prophets. 7. And Saul said to his servants, "Seek a woman who is a mistress of a spirit for me, so that I may go to her and inquire through her."

And his servants said to him, "Here's a woman who is a mistress of a spirit in En-dor." 8. And Saul disguised himself and wore other clothes and went, he and two people with him, and they came to the woman by night.

And he said, "Communicate through a spirit for me, and bring up the one whom I'll tell you."

9. And the woman said to him, "Here, you know what Saul has done, that he cut off the mediums and necromancers from the land, and why do you jeopardize my life, to kill me?"

10. And Saul swore to her by YHWH, saying, "As YHWH lives, no crime will be attributed to you in this matter."

11. And the woman said, "Whom shall I bring up for you?"

And he said, "Bring up Samuel for me."

12. And the woman saw Samuel and cried out in a loud voice, and the woman said to Saul, saying, "Why did you deceive me? You're Saul!"

13. And the king said to her, "Don't be afraid. But what did you see?"

And the woman said to Saul, "I saw gods coming up from the earth!"

14. And he said to her, "What is its appearance?"

And she said, "An old man coming up. And he's wrapped in a robe." And Saul knew that it was Samuel. And he knelt, nose to the ground, and bowed.

15. And Samuel said to Saul, "Why have you disturbed me, bringing me up?"

And Saul said, "I'm very distressed, because the Philistines are making war against me, and God has turned away from me and won't answer me anymore, neither by the hand of prophets nor through dreams, and so I've called to you to let me know what I should do."

16. And Samuel said, "And why do you ask me, since YHWH has turned away from you and has become your opponent? 17. And YHWH has brought about for Himself the thing that He spoke by my hand, and YHWH has torn the kingdom from your hand and given it to one of your fellow men, to David, 18. because you didn't listen to YHWH's voice and didn't carry out his anger against Amalek. That is why YHWH has done this thing to you this day. 19. And YHWH will also deliver Israel with you into the Philistines' hand. And tomorrow you and your sons will be with me. Also, YHWH will deliver the camp of Israel into the Philistines' hand."

20. And Saul hurried and fell, his full length, to the ground and was very afraid from Samuel's words. There was also no strength in him because he had not eaten bread all that day and all that night. 21. And the woman came to Saul and saw that he was very terrified, and she said to him, "Here, your maid listened to your voice, and I took my life in my hand and listened to your words that you spoke to me. 22. And now, listen, you as well, to your maid's voice, and let me set a piece of bread in front of you, and eat, and there will be strength in you when you go on the road."

23. And he refused and said, "I won't eat." And his servants and the woman, too, pressed him, and he listened to their voice and got up from the ground and sat on the bed. 24. And the woman had a fattened calf in the house, and she hurried and slaughtered it and took flour and kneaded it and baked unleavened bread with it. 25. And she brought it out it in front of Saul and in front of his servants, and they ate. And they got up and went in that night.

29:1. And the Philistines gathered all their camps to Aphek, and Israel was camped at the spring that is in Jezreel. 2. And the lords of the Philistines had been passing on by hundreds and by thousands, and David and his men passing last with Achish. 3. And the officers of the Philistines said, "What are these Hebrews?"

And Achish said to the officers of the Philistines, "Isn't this David, servant of Saul, king of Israel, who has been with me for days now, or for *years* now, and I haven't found anything against him from the day he fell away until this day."

4. And the officers of the Philistines were angry at him. And the lords of the Philistines said to him, "Send the man back, and let him go back to his place that you've allocated him, and let him not go down into war with us so he won't be an adversary for us in the war; for by what means would this one be reconciled with his lord: wouldn't it be by the heads of those men?! 5. Isn't this David of whom they sang in the dances, saying, 'Saul has struck his thousands and David his ten thousands'?"

6. And Achish called David and said to him, "As YHWH lives, you're honest, and your going and coming with me in the camp is good in my eyes because I haven't found anything bad in you from the day you came to me to this day, but you're not good in the lords' eyes. 7. And now, go back and go in peace so you won't do bad in the eyes of the lords of the Philistines."

8. And David said to Achish, "But what have I done, and what have you found in your servant from the day that I was in front of you to this day that I shouldn't come and fight against the enemies of my lord the king?"

9. And Achish answered, and he said to David, "I know that you're as good in my eyes as an angel of God, but the officers of the Philistines have said, 'He won't come up with us into the war.' 10. And now, get up early in the morning, you and your lord's servants who came with you. And you'll get up early in the morning and have light, and go."

11. And David got up early, he and his men, to go in the morning, to go back to the land of the Philistines, and the Philistines went up to Jezreel.

30:1. And it was when David and his men came to Ziklag on the third day: and the Amalekites had raided the Negeb and Ziklag. And they struck Ziklag and burned it in fire 2. and captured the women who were in it, from the youngest to the oldest. They did not kill anyone, and they carried them away and went their way. 3. And David and his men came to the city; and, here, it was burned in fire, and their wives and their sons and their daughters were taken captive. 4. And David and the people who were with him raised their voice and wept until there was no strength in them to weep. 5. And David's two wives had been taken captive, Ahinoam, the Jezreelite, and Abigail, Nabal's wife, the Carmelite. 6. And David was very distressed because the people said to stone him, because all the people's soul was bitter, each for his sons and his daughters. And David fortified himself through YHWH, his God. 8. And David asked YHWH, saying, "If I pursue this band, shall I overtake them?"

And He said to him, "Pursue, because you'll *overtake* and you'll *rescue.*"

9. And David went, he and six hundred men who were with him, and they came to the wadi Besor, and some who were left behind stayed; 10. and David pursued, he and four hundred men, and two hundred men who were too exhausted to cross the wadi Besor stayed.

11. And they found an Egyptian man in a field and took him to David, and they gave him bread, and he ate, and they gave him water to drink, 12. and they gave him a slice of fig cake and two

raisin clusters. And he ate, and his spirit came back to him, because he had not eaten bread and had not drunk water three days and three nights. 13. And David said to him, "To whom do you belong, and from where do you come?"

And he said, "I'm an Egyptian boy, an Amalekite man's servant, and my lord left me because I was sick three days ago. 14. We raided the Negeb of the Cheretites and the part that is Judah's and the Negeb of Caleb, and we burned Ziklag in fire."

15. And David said to him, "Will you take me down to this band?"

And he said, "Swear to me by God that you won't kill me and won't shut me up in my lord's hand, and I'll take you down to this band."

16. And he took him down, and here they were, spread out all over the ground, eating and drinking and celebrating with all the big spoil that they had taken from the Philistines' land and from the land of Judah. 17. And David struck them from the twilight to the evening of the following day, and not a man of them escaped except for four hundred young men who rode on camels and fled. 18. And David rescued everyone whom Amalek had taken, and David rescued his two wives, 19. and they had nothing lacking, from the youngest to the oldest and to boys and girls, and from the spoil to everything that they had taken: David brought back everything. 20. And David took the sheep and the oxen—they drove them in front of the cattle—and they said, "This is David's spoil." 21. And David came to the two hundred men who had been too exhausted to go after David, whom they had made to stay at the wadi Besor, and they came out to David and to the people who were with him. And David went over to the people and asked them how they were.

22. And every bad and good-for-nothing man among the men who went after David answered, and they said, "Since they didn't

go up with me, we won't give them any of the spoil that we rescued except for each man his wife and children, and let them drive off and go."

23. And David said, "You shall not do this, my brothers, because YHWH gave it to us, and He watched over us and delivered the band that came against us into our hand. 24. And who will listen to you about this thing? Rather, the portion of the one who went down into battle and the portion of the one who sat with the equipment will be the same: they'll share together." 25. And it was from that day forward, and he made it a statute and law for Israel to this day.

26. And David came to Ziklag and sent some of the spoil to the elders of Judah, to his compatriot tribe, saying, "Here's a blessing for you from the spoil of YHWH's enemies," 27. for those in Beth-El and for those in Ramoth-Negeb and for those in Jattir 28. and for those in Aroer and for those in Siphmoth and for those in Eshtemoa 29. and for those in Rachal and for those in the cities of the Jerahmeelite and for those in the cities of the Kenite 30. and for those in Hormah and for those in Bor-ashan and for those in Athach 31. and for those in Hebron, and for all the places in which David went, he and his men.

31:1. And the Philistines were fighting against Israel, and the men of Israel fled from the Philistines, and they fell, dead bodies, in Mount Gilboa. 8. And it was on the following day, and the Philistines came to strip the dead bodies, and they found Saul and his three sons fallen in Mount Gilboa. 9. And they cut off his head and stripped his armor and sent word all around the land of the Philistines to give the news at the temple of their idols and to the people. 10. And they put his armor in the temple of Ashtaroth, and they pinned his body to the wall of Beth-shan. 11. And the people who lived in Jabesh-Gilead heard what the Philistines had done to Saul. 12. And all the warriors got up and went all night and took Saul's body and his sons' bodies from the wall of Beth-shan and came to Jabesh and burned them there, 13. and they took their bones and buried them under the tamarisk tree in Jabesh, and they fasted seven days.

2 Samuel 1:1. And it was after Saul's death, when David had come back from striking Amalek, and David stayed in Ziklag two days. 2. And it was on the third day, and here was a man coming from the camp, from having been with Saul, and his clothes were torn, and dirt was on his head. And it was as he came to David, and he fell to the ground and bowed.

3. And David said to him, "Where are you coming from?"

And he said to him, "I escaped from the camp of Israel."

4. And David said to him, "What was the matter? Tell me."

And he said, "That the people fled from the battle, and also many of the people fell and died, and also Saul and Jonathan, his son, died."

5. And David said to the young man who was telling him, "How do you know that Saul and Jonathan, his son, died?"

6. And the young man who was telling him said, "I *happened* to be in Mount Gilboa, and here was Saul leaning on his spear; and, here, the chariots and horsemen had caught up to him. 7. And he looked behind him and saw me, and he called to me, and I said, 'I'm here.' 8. And he said to me, 'Who are you?' And I said to him, 'I'm an Amalekite.' 9. And he said to me, 'Stand over me and kill me, because pangs have seized me for as long as my life is still in me.' 10. And I stood over him and killed him because I knew that he wouldn't live after his fall. And I took the crown that was on his head and the bracelet that was on his arm and brought them to my lord here."

11. And David took hold of his clothes and ripped them—and all the people who were with him as well—12. and they mourned and wept and fasted until evening for Saul and for Jonathan, his son, and for YHWH's people and for the house of Israel, because they had fallen by the sword.

13. And David said to the young man who had told him, "Where do you come from?"

And he said, "I'm the son of an Amalekite resident."

14. And David said to him, "How were you not afraid to put out your hand to destroy YHWH's anointed?" 15. And David called one of the young men and said, "Go over. Strike him." And he struck him, and he died. 16. And David said to him, "Your blood is on your head, because your own mouth testified against you, saying, 'I killed YHWH's anointed.'"

17. And David made this lament over Saul and over Jonathan, his son, 18. and he said to teach it to the children of Judah. Here, it is written in the book of Yashar:

19. The gazelle, Israel, is slain on your high places.
How have the mighty fallen!

20. Don't tell it in Gath,
don't give the news in Ashkelon's streets,
or else the Philistines' daughters will be happy,
or else the uncircumcised's daughters will rejoice.

21. Mountains of Gilboa, let there be no dew,
and let there be no rain on you, and fields of offerings,
for a shield of warriors was rejected there,
Saul's shield, unanointed with oil.

22. From blood of those slain,
from fat of warriors,

Jonathan's bow didn't turn back,
and Saul's sword didn't come back empty.

23. Saul and Jonathan,
beloved and pleasing in their lives
and not separated in their death,
they were swifter than eagles,
they were stronger than lions.

24. Daughters of Israel, weep for Saul,
who dressed you in scarlet with delights,
who put golden ornaments on your clothes.

25. How have the mighty fallen in war!
Jonathan, slain on your high places,

26. I'm distressed over you, my brother, Jonathan.
You were very pleasant to me.
Your love was more wondrous to me
than love of women.

27. How have the mighty fallen
and weapons of war perished.

2:1. And it was after this, and David asked of YHWH, saying, "Shall I go up into one of the cities of Judah?"

And YHWH said to him, "Go up."

And David said, "Where shall I go up?"

And He said, "To Hebron."

2. And David went up there, and his two wives as well, Ahinoam, the Jezreelite, and Abigail, Nabal's wife, the Carmelite. 3. And David brought up the people who were with him, each man and his house, and they lived in the cities of Hebron. 4. And the people of Judah came and anointed David there as king over the house of Judah.

And they told David, saying, "It was the people of Jabesh-Gilead who buried Saul." 5. And David sent messengers to the people of Jabesh-Gilead and said to them, "You're blessed to YHWH because you did this kindness toward your lord, toward Saul, and buried him. 6. And now, may YHWH practice kindness and faithfulness with you, and I'll practice such good toward you as well, because you did this thing. 7. And now, let your hands be strong, and be warriors, because your lord, Saul, is dead, and also the house of Judah have anointed me as king over them."

8. And Abner, son of Ner, who had been Saul's army commander, had taken Ishbaal, son of Saul, and brought him over to Mahanaim, 9. and they made him king over Gilead and over the Ashurites and over Jezreel and over Ephraim and over Benjamin and over the whole of Israel. 10b. But the house of Judah were behind David.

12. And Abner, son of Ner, and the servants of Ishbaal, son of Saul, went out, from Mahanaim to Gibeon. 13. And Joab, son of Zeruiah, and David's servants went out. And they met them by the pool of Gibeon together and sat: these on one side of the pool and those on the other side of the pool. 14. And Abner said to Joab, "Let the boys get up and perform for us!"

And Joab said, "Let them get up!"

15. And they got up and passed over by number—twelve of Benjamin and of Ishbaal, son of Saul, and twelve from David's servants—16. and they took hold, each man at his opponent's head and his sword in his opponent's side, and fell together. And that place was called Helkath-hazzurim, which is in Gibeon. 17. And the battle was extremely hard on that day, and Abner and the people of Israel were stricken before David's servants.

18. And the three sons of Zeruiah were there: Joab and Abishai and Asahel. And Asahel was as light on his feet as one of the gazelles that are in the field. 19. And Asahel pursued Abner, and he did not turn to go to the right or to the left from behind Abner. 20. And Abner looked behind him and said, "Is this you, Asahel?"

And he said, "I am."

21. And Abner said to him, "Turn to your right or to your left and seize one of the boys and take what can be pulled off him for yourself." But Asahel was not willing to turn away from behind him. 22. And Abner went on again to say to Asahel, "Turn away from behind me. Why should I strike you to the ground? And how could I hold up my face to Joab, your brother?" 23. And he refused to turn away. And Abner struck him with the back end of the spear in the abdomen, and the spear came out behind him, and he fell there and died at the spot. And everyone who came to the place where Asahel fell and died stopped, 24. but Joab and Abishai pursued Abner. And the sun was setting, and they came to the hill of Ammah, which faces Giah, on the way of the wilderness of Gibeon. 25. And the children of Benjamin gathered behind Abner and became one troop and stood on top of one hill. 26. And Abner called to Joab and said, "Will the sword devour forever? Don't you know that it will be bitter in the end? And how long will you not say to the people to turn back from going after their brothers?"

27. And Joab said, "As God lives, if you hadn't spoken, then the people would have been suspended, each man, from going after his brother, this morning! 28. And Joab blasted on a ram's horn, and all the people stopped and did not pursue Israel anymore, and they did not go on fighting anymore. 29. And Abner and his men went in the Arabah all that night and crossed the Jordan and went through all of Bithron and came to Mahanaim. 30. And Joab turned back from going after Abner and gathered all the people.

And nineteen men and Asahel were missing out of David's servants; 31. and David's servants struck dead three hundred sixty out of Benjamin and out of Abner's men. 32. And they carried Asahel and buried him in his father's tomb, which was at Bethlehem, and Joab and his men went all night, and the day dawned on them at Hebron.

3:1. And the war was long between the house of Saul and the house of David, and David went on getting stronger, and the house of Saul went on getting weaker.

2. And children were born to David in Hebron: And his firstborn was Amnon, to Ahinoam the Jezreelite. 3. And his second was Chileab, to Abigail, Nabal's wife, the Carmelite. And the third was Absalom, son of Maachah, the daughter of Talmai, king of Geshur. 4. And the fourth was Adonijah, son of Haggith. And the fifth was Shephatiah, son of Abital. 5. And the sixth was Ithream, by Eglah, David's wife. These were born to David in Hebron.

6. And it was while the war was going on between the house of Saul and the house of David: and Abner was becoming powerful in the house of Saul. 7. And Saul had had a concubine, and her name was Rizpah, daughter of Aiah; and Ishbaal said to Abner, "Why did you come to my father's concubine?"

8. And Abner was very angry over Ishbaal's words, and he said, "Am I a dog's head belonging to Judah? Today I'm practicing kindness with the house of Saul, your father, toward his brothers and toward his friends, in that I haven't delivered you into David's hand! And you charge me with a crime about the woman today?! 9. May God do this to Abner and double this to him if I don't do for David what YHWH swore to him: 10. to cause the kingdom to pass

from the house of Saul and to set up David's throne over Israel and over Judah, from Dan to Beer-sheba!"

11. And he was not able to answer Abner another word because of his fear of him.

12. And Abner sent messengers to David on his behalf, saying, "Whose is the land?" and saying, "Make your covenant with me, and my hand will be with you here to bring all of Israel around to you."

13. And he said, "Good. I'll make a covenant with you; just one thing I ask from you, saying, 'You won't see my face unless you bring Michal, daughter of Saul, when you come to see my face.'"

14. And David sent messengers to Ishbaal, son of Saul, saying, "Give me my wife, Michal, whom I betrothed to me by a hundred foreskins of Philistines."

15. And Ishbaal sent and took her from the man, from Paltiel, son of Laish. 16. And her man went with her, going on weeping behind her, up to Bahurim. And Abner said, "Go. Go back." And he went back.

17. And Abner's word was conveyed to Israel's elders, saying, "Both yesterday and the day before, you were seeking David as king over you. 18. And now, do it. Because YHWH has said of David, saying, 'By the hand of David, my servant, I'll save my people, Israel, from the Philistines' hand and from all their enemies' hand.'" 19. And Abner also spoke in the ears of Benjamin. And Abner also went to speak in David's ears in Hebron everything that was good in Israel's eyes and in the eyes of all the house of Benjamin. 20. And Abner came to David in Hebron, and twenty people with him, and David made a feast for Abner and for the people who were with him. 21. And Abner said to David, "Let me get up and go and gather all of Israel to my lord the king, and

they'll make a covenant with you, and you'll rule over all that your soul desires." And David let Abner go, and he went in peace.

22. And here were David's servants and Joab coming from a raid, and they brought great spoil with them. And Abner was not with David in Hebron, because he let him go, and he went in peace. 23. And Joab and all the army that was with him had come, and they told Joab, saying, "Abner, son of Ner, came to the king, and he let him go, and he went in peace."

24. And Joab came to the king and said, "What have you done? Here, Abner came to you. Why did you let him go? And he's gone off. 25. You know Abner, son of Ner, that he came to trick you and to know your exits and entrances and to know everything that you're doing!" 26. And Joab came out from David and sent messengers after Abner, and they brought him back from the cistern of Sirah. And David did not know. 27. And Abner came back to Hebron, and Joab took him aside inside the gate to speak with him in private, and he struck him there in the abdomen, and he died over the blood of Asahel, his brother.

28. And David heard afterwards, and he said, "I and my kingdom are innocent of the blood of Abner, son of Ner, before YHWH forever. 29. Let it whirl around Joab's head and on all of his father's house, and let there not be lacking from Joab's house someone with gonorrhea and someone with skin disease and someone who holds a spindle and someone who falls by the sword and someone who lacks bread." 30. And Joab and Abishai, his brother, had killed Abner because he killed Asahel, their brother, in Gibeon in battle. 31. And David said to Joab and to all the people who were with him, "Tear your clothes and wear sacks and mourn in front of Abner." And King David was going behind the bier. 32. And they buried Abner in Hebron, and the king raised his voice and wept at Abner's tomb, and all the people wept. 33. And the king lamented Abner and said,

Should Abner die like the death of a fool?
34. Your hands weren't bound by manacles,
and your feet weren't put in shackles.
You fell the way one falls before malicious people.

And all the people continued to weep over him. 35. And all the people came to get David to eat bread while it was still daytime, but David swore, saying, "May God do this to me and double this if I taste bread or anything at all before the sun sets. 36. And all the people recognized, and it was good in their eyes, as everything that the king did was good in all the people's eyes, 37. and all the people and all of Israel knew on that day that killing Abner, son of Ner, was not from the king. 38. And the king said to his servants, "Don't you know that an officer and a big man has fallen today in Israel? 39. And I'm weak today, though anointed king; and these men, the sons of Zeruiah, are harder than I. Let YHWH pay back the one who does bad according to his bad act."

4:1. And Saul's son heard that Abner had died in Hebron, and his hands were weak. And all of Israel were terrified. 2. And Saul's son had two men who were officers of troops; one's name was Baanah, and the second one's name was Rechab, sons of Rimmon, the Beerothite, from the children of Benjamin (because Beeroth was reckoned to Benjamin as well; 3. and the Beerothites had fled to Gittaim and have been residing there to this day). 4. And Jonathan, son of Saul, had a son with injured legs. He was five years old when news of Saul and Jonathan had come from Jezreel, and his nurse picked him up and fled, and it was in her hurry to flee that he fell and was crippled. And his name was Mephibaal. 5. And the sons of Rimmon, the Beerothite, Rechab and Baanah, went, and they came to Ishbaal's house in the heat of the day. And he was lying down at noon, 6. and here they came inside the house as if getting wheat, and they struck him to the abdomen, and Rechab and Baanah, his brother, escaped. 7. And they had come to the house when he was lying on his bed in his bedroom, and they killed him and cut off his head and took his head and went by way of the Arabah all night. 8. And they brought Ishbaal's head to David at Hebron and said to the king, "Here's the head of Ishbaal, son of Saul, your enemy, who sought your life, and YHWH has given my lord the king revenge this day from Saul and from his seed!"

9. And David answered Rechab and Baanah, his brother, the sons of Rimmon, the Beerothite, and he said to them, "As YHWH lives, who has redeemed my soul from every trouble: 10. when someone told me, saying, 'Here, Saul has died'—and in his eyes he was like someone bringing good news—I took hold of him and killed him in Ziklag, the one who expected me to give him something for the news. 11. How much more when wicked men have killed a virtuous man in his house on his bed. And now, won't I demand his blood from your hand? And I'll burn you out from the land!" 12. And David commanded the young men, and they killed them and cut off their hands and their feet and hung them over the pool in Hebron. And they took Ishbaal's head and buried it in Abner's tomb in Hebron.

5:1. And all the tribes of Israel came to David at Hebron and said, saying, "Here we're your bone and your flesh. 2. Both yesterday and the day before, when Saul was king over us, *you* were the one who brought Israel out and in; and YHWH has said to you, '*You* shall shepherd my people, Israel, and *you* shall become the designated king over Israel.'" 3. And all of Israel's elders came to the king at Hebron, and King David made them a covenant in Hebron before YHWH, and they anointed David as king over Israel. 4. David was thirty years old when he reigned; he reigned forty years: 5. in Hebron he reigned over Judah seven years and six months, and in Jerusalem he ruled thirty-three years over all Israel and Judah.

6. And the king and his people went to Jerusalem, to the Jebusites who lived in the land. And they said to David, saying, "You won't come here, because the blind and the crippled could turn you away," saying, "David won't come here!" 7. But David captured the stronghold of Zion. It is the City of David.

8. And David said on that day, "Anyone who would strike the Jebusites, let him get there through the water shaft—and the crippled and the blind who are hated by David's soul." On account of this they say, "The blind and crippled shall not come into the house." 9. And David lived in the fortress and called it "The City of David." And David built all around from the Millo and inward.

10. And David went on getting bigger, and YHWH, God of hosts, was with him.

11. And Hiram, king of Tyre, sent messengers to David, and cedar trees and carpenters and masons, and they built a house for David. 12. And David knew that YHWH had established him as king over Israel and that He had supported his kingdom for the sake of His people, Israel.

13. And David took more concubines and wives from Jerusalem after he came from Hebron, and more sons and daughters were born to David. 14. And these are the names of the ones who were born to him in Jerusalem: Shammua and Shobab and Nathan and Solomon 15. and Ibhar and Elishua and Nepheg and Japhia 16. and Elishama and Eliada and Eliphalet.

17. And the Philistines heard that they had anointed David as king over Israel, and all the Philistines went up to seek David, and David heard and went down to the fortress. 18. And the Philistines had come, and they spread out in the valley of Rephaim. 19. And David asked YHWH, saying, "Shall I go up to the Philistines? Will you deliver them into my hand?"

And YHWH said to David, "Go up, because I'll *deliver* the Philistines into your hand!"

20. And David came to Baal-perazim, and David struck them there and said, "YHWH has burst through my enemies before me like a water burst." On account of this the name of that place was called Baal-perazim. 21. And they left their idols there, and David and his men carried them away.

22. And the Philistines proceeded to come up again and spread out in the valley of Rephaim. 23. And David asked YHWH, and He said, "You shall not go up. Go around behind them and come to them across from the balsam trees. 24. And it will be when you hear the sound of marching in the tops of the balsam trees, then you'll move, because then YHWH will have gone out ahead of you to strike the Philistines' camp." 25. And David did so, as YHWH had commanded him, and he struck the Philistines from Geba to where you come to Gezer.

6:1. And David again gathered all the chosen men in Israel, thirty thousand. 2. And David and all the people who were with him got up and went from Baalê-Judah to bring up from there God's ark, which is called "Name": *"The Name of YHWH of Hosts Who Is Enthroned on the Cherubs."* 3. And they had God's ark ride on a new wagon and carried it from Abinadab's house, which was in Gibeah; and Uzzah and Ahio, Abinadab's sons, were driving the new wagon. 4. And they carried it from Abinadab's house, which was in Gibeah, with God's ark, and Ahio was going in front of the ark. 5. And David and all the house of Israel were performing before YHWH with all their might and with songs and lyres and harps and drums and rattles and cymbals.

6. And they came to the threshing floor of Nachon. And Uzzah reached out to God's ark and held on to it because the oxen faltered.

7. And YHWH's anger flared at Uzzah, and God struck him there because he had reached out his hand on the ark, and he died there by God's ark. 8. And David was angry that YHWH had burst out at Uzzah, and he called that place Perez-Uzzah to this day.

9. And David feared YHWH on that day, and he said, "How can YHWH's ark come to me?" 10. And David was not willing to turn YHWH's ark to him at the City of David, and David directed it to the house of Obed-edom, the Gittite, 11. and YHWH's ark stayed at the house of Obed-edom, the Gittite, three months. And YHWH blessed Obed-edom and all of his house. 12. And it was told to King David, saying, "YHWH has blessed Obed-edom's house and everything that he has on account of God's ark." And David went and brought up God's ark from Obed-edom's house to the City of David with happiness. 13. And it was, when those who were carrying the ark had marched six steps, that he sacrificed an ox and a fatling. 14. And David was whirling with all his might before YHWH, and David was wearing a linen ephod. 15. And David and all the house of Israel were bringing up YHWH's ark with shouting and the sound of the ram's horn. 16. And YHWH's ark was coming to the City of David, and Michal, Saul's daughter, gazed through the window and saw King David leaping and whirling before YHWH, and she disdained him in her heart.

17. And they brought YHWH's ark and set it up in its place inside the tent that David had pitched for it, and David made burnt offerings and peace offerings before YHWH. 18. And David finished from making the burnt offering and the peace offerings, and he blessed the people in the name of YHWH of hosts. 19. And he distributed to all the people, to all the mass of Israel, from man to woman, to each one, one loaf of bread and one cake and one raisin-cake; and all the people went, each one to his house.

20. And David went back to bless his house. And Michal, Saul's daughter, came out to David and said, "How honored today was the king of Israel, who was exposed today to the eyes of his servants' maids the way one of the worthless men would *expose himself!*"

21. And David said to Michal, "It was before YHWH, who chose me over your father! And over all of his house! And commanded that I be the designated king over YHWH's people, over Israel! And I'll perform before YHWH, 22. and I'll be even more inconsequential than this, and I'll be low in my eyes and with the maids with whom you said I was honored!"

23. And Michal, Saul's daughter, had no child to the day of her death.

7:1a. And it was when the king lived in his house: 2. and the king said to Nathan, the prophet, "Look, I live in a cedar house, while God's ark stays inside curtains."

3. And Nathan said to the king, "Go, do everything that's in your heart, because YHWH is with you."

4. And it was on that night, and YHWH's word came to Nathan, saying, 5. "Go and say to my servant, to David, 'YHWH said this: Will *you* build *me* a house for me to stay in? 6. Because I haven't stayed in a house from the day I brought the children of Israel up from Egypt to this day, but I've been going about in a tent and in a tabernacle. 7. Wherever I've gone, among all the children of Israel, did I speak a word with one of Israel's tribes whom I commanded to shepherd my people, Israel, saying, "Why haven't you built me a cedar house?"' 8. And now, say this to my servant, David: 'YHWH of hosts said this: I took you from the pasture, from behind the sheep, to be the designated king over my people,

over Israel, 9. and I was with you wherever you went, and I cut off all your enemies before you. And I'll make a big name for you, like the name of the biggest in the earth, 10. and I'll set a place for my people, for Israel, and plant them, and they'll settle in their place and not be distressed anymore, and unjust men will never degrade them again as before, 11a. from the day that I commanded judges to be over my people, Israel.' 11c. And YHWH has told you that YHWH will make a house for *you:* 12. 'When your days will be fulfilled and you'll lie with your fathers, I'll set up your seed after you, one who will come out of your insides, and I'll make his kingdom secure.'"

18. And King David came and sat before YHWH and said, "Who am I, my Lord YHWH, and who is my house, that you've brought me this far? 19. And even this was a small thing in your eyes, my Lord YHWH, because you've spoken about your servant's house for a distant time as well; and this is instruction to a human, my Lord YHWH. 20. And what more can David add to speak to you, since you know your servant, my Lord YHWH. 21. For the sake of your word and according to your heart you've done all of this big thing, to cause your servant to know. 25. And now, YHWH, God, bring about the thing that you spoke about your servant and about his house forever, and do as you have spoken. 26. And let your name be big forever, saying, 'YHWH of hosts is God over Israel,' and let the house of your servant, David, be established before you. 27. Because you, YHWH of hosts, God of Israel, have disclosed in your servant's ear, saying, 'I'll build a house for you.' On account of this, your servant found it in his heart to make this prayer to you. 28. And now, my Lord YHWH, you are God, and your words will be true, and you spoke this good thing to your servant. 29. And now, be pleased, and bless your servant's house to be before you forever, because you, my Lord YHWH, have

spoken, and may your servant's house be blessed with your blessing forever."

8:1. And it was after this, and David struck the Philistines and subdued them, and David took Metheg-ammah from the Philistines' hand.

2. And he struck Moab and measured them with a rope, making them lie down on the ground, and he measured two rope lengths to put to death and one full length of the rope to keep alive. And Moab became tribute-bearing servants to David.

3. And David struck Hadadezer, son of Rehob, king of Zobah, when he went to bring his power back to the Euphrates river. 4. And David captured one thousand seven hundred horsemen and twenty thousand foot soldiers of his, and David hamstrung all the chariot horses but left a hundred chariot horses of them.

5. And Aram of Damascus came to help Hadadezer, king of Zobah, and David struck twenty-two thousand men of Aram. 6. And David put garrisons in Aram of Damascus, and Aram became tribute-bearing servants to David. And YHWH made David victorious wherever he went. 7. And David took the golden shields that Hadadezer's servants had and brought them to Jerusalem, 8. and King David took a very great quantity of bronze from Betah and Berothai, cities of Hadadezer.

9. And Toi, king of Hamath, heard that David had struck Hadadezer's whole army, 10. and Toi sent Joram, his son, to King David to ask how he was and to bless him because he had fought against Hadadezer and struck him, because Hadadezer had been Toi's enemy in wars, and in his hand were objects of silver and objects of gold and objects of bronze. 11. King David also consecrated them to YHWH with the silver and the gold that he had

consecrated from all the nations that he had conquered, 12. from Aram and from Moab and from the children of Ammon and from the Philistines and from Amalek and from the spoil of Hadadezer, son of Rehob, king of Zobah.

13. And when he came back, David made a name for himself by his striking Edom in the Valley of Salt, eighteen thousand. 14. And he put garrisons in Edom—in all of Edom he put garrisons—and all of Edom were servants to David. And YHWH saved David wherever he went.

15. And David reigned over all Israel, and David was executing judgment and justice to all his people. 16. And Joab, son of Zeruiah, was over the army; and Jehoshaphat, son of Ahilud, was recorder; 17. and Zadok, son of Ahitub, and Ahimelech, son of Abiathar, were priests; and Seraiah was scribe; 18. and Benaiah, son of Jehoiada, was over the Cherethites and the Pelethites; and David's sons were priests.

9:1. And David said, "Is there still anyone who is left from the house of Saul, so I can practice kindness with him for Jonathan's sake?" 2. And there was a servant from the house of Saul, and his name was Ziba, and they called him to David. And the king said to him, "Are you Ziba?"

And he said, "Your servant."

3. And the king said, "Isn't there still any man from the house of Saul, so I can practice God's kindness with him?"

And Ziba said to the king, "There's still a son of Jonathan, with injured legs."

4. And the king said to him, "Where is he?"

And Ziba said to the king, "He's here at the house of Machir, son of Ammiel, in Lo-debar."

5. And King David sent and took him from the house of Machir, son of Ammiel, from Lo-debar. 6. And Mephibaal, son of Jonathan, son of Saul, came to David and fell on his face and bowed. And David said, "Mephibaal."

And he said, "Your servant is here."

7. And David said to him, "Don't be afraid, because I'll practice kindness with you for the sake of Jonathan, your father. And I'll give back to you all the property of Saul, your father. And you'll eat bread at my table always."

8. And he bowed and said, "What is your servant that you've turned to a dead dog like me?"

9. And the king called Ziba, Saul's young man, and said to him, "I've given everything that Saul had and all of his house had to your lord's son. 10. And you'll work the land for him, you and your children and your servants, and bring it, and your lord's son will have food and eat it; and Mephibaal, your lord's son, will eat bread at my table always."

And Ziba had fifteen sons and twenty servants. 11. And Ziba said to the king, "According to everything that my lord the king has commanded his servant, so your servant will do." And so Mephibaal was eating at David's table like one of the king's sons. 12. And Mephibaal had a small son, and his name was Micah. And all who lived in Ziba's house were servants to Mephibaal. 13. And so Mephibaal was living in Jerusalem, because he was eating at the king's table always, and he was crippled in both of his legs.

10:1. And it was after this, and the king of the children of Ammon died, and his son, Hanun, ruled in his place. 2. And David said, "Let me practice kindness with Hanun, son of Nahash, as his father practiced kindness with me." And David sent by his servants' hands to console him over his father. And David's servants came to the land of the children of Ammon.

3. And the officers of the children of Ammon said to Hanun, their lord, "In your eyes, is David honoring your father because he sent consolers? Isn't it for the purpose of searching the city and to spy it out and to overturn it that David sent his servants to you?"

4. And Hanun took David's servants and shaved off half of their beards and cut their clothes in half, to their buttocks, and sent them away. 5. And they told David, and he sent to them because the men were very humiliated, and the king said, "Stay in Jericho until your beards grow, and then come back."

6. And the children of Ammon saw that they had been made odious to David, and the children of Ammon sent and hired Aram of Beth-rehob and Aram of Zobah, twenty thousand foot soldiers; and the king of Maacah, a thousand men; and men of Tob, twelve thousand men. 7. And David heard, and he sent Joab and all the army of the powerful warriors. 8. And the children of Ammon came out and aligned for war at the entrance of the gate; and Aram of Zobah and Rehob and men of Tob and Maacah were by themselves in the field. 9. And Joab saw that the battlefront was against him from the front and from the rear. And he chose some from all of the chosen men of Israel and aligned toward Aram 10. and placed the rest of the people in the hand of Abishai, his brother, and he aligned toward the children of Ammon. 11. And he said, "If Aram will be too strong for me, then you'll be my salvation; and if the children of Ammon will be too strong for you, then I'll go to save you. 12. Be strong, and let's strengthen ourselves for our people and for our God's cities, and YHWH will do what is good in His eyes." 13. And Joab and the people who were with him went over

to the battle against Aram, and they fled from him. 14. And the children of Ammon saw that Aram had fled, and they fled from Abishai and came into the city. And Joab came back from the children of Ammon and came to Jerusalem.

15. And Aram saw that it was stricken before Israel, and they were gathered together. 16. And Hadadezer sent and brought out Aram, who were from the other side of the river, and they came to Helam; and Shobach, the commander of Hadadezer's army, was in front of them. 17. And it was told to David, and he gathered all of Israel and crossed the Jordan and came to Helam. And Aram aligned toward David and fought with him. 18. And Aram fled from Israel, and David killed seven hundred chariots and forty thousand horsemen from Aram, and he struck Shobach, the commander of its army, and he died there. 19. And all the kings, the servants of Hadadezer, saw that they were stricken before Israel, and they made peace with Israel and served them, and Aram were afraid to save the children of Ammon anymore.

11:1. And it was at the turning of the year, at the time when the kings go out, and David sent Joab and his servants with him and all of Israel, and they destroyed the children of Ammon and besieged Rabbah. And David was staying in Jerusalem. 2. And it was at evening time, and David got up from his bed and walked on the roof of the king's house and saw a woman washing from the roof, and the woman was very good-looking. 3. And David sent and inquired about the woman, and one said, "Isn't this Bathsheba, daughter of Eliam, wife of Uriah, the Hittite?" 4. And David sent messengers and took her, and she came to him, and he lay with her—and she was cleansing herself from her impurity—and she went back to her house. 5. And the woman became pregnant, and she sent and told David. And she said, "I'm pregnant."

6. And David sent to Joab, "Send Uriah, the Hittite, to me."

And Joab sent Uriah to David. 7. And Uriah came to him, and David asked how Joab was and how the people were and how the war was going. 8. And David said to Uriah, "Go down to your house and wash your feet." And Uriah went out from the king's house, and a portion from the king went out behind him. 9. And Uriah lay at the entrance of the king's house with all of his lord's servants, and he did not go down to his house.

10. And they told David, saying, "Uriah didn't go down to his house."

And David said to Uriah, "Haven't you come a distance? Why didn't you go down to your house?"

11. And Uriah said to David, "The ark and Israel and Judah are living in shelters, and my lord, Joab, and my lord's servants are camping in the open field; and should *I* come to my house to eat and drink and lie with my wife?! As you live, and as your soul lives, I won't do this thing!"

12. And David said to Uriah, "Stay here today, too, and I'll let you go tomorrow." And Uriah stayed in Jerusalem on that day and on the following. 13. And David called him, and he ate in front of him. And he drank. And he got him drunk. And in the evening he went out to lie down in his bed with his lord's servants, and he did not go down to his house.

14. And it was in the morning, and David wrote a scroll to Joab and sent it by Uriah's hand. 15. And he wrote in the scroll, saying, "Set Uriah facing the battle's strongest front—and you shall fall back from behind him so he'll be struck and die." 16. And it was when Joab was watching the city: and he put Uriah at a place

where he knew that there were warriors. 17. And the people of the city came out and fought Joab, and some of the people fell, some of David's servants, and Uriah, the Hittite, died as well.

18. And Joab sent and told David all the details of the battle, 19. and he commanded the messenger, saying, "When you finish speaking all the details of the battle to the king, 20. if the king's fury rises, and he says to you, 'Why did you come up to the city to fight? Didn't you know that they'd shoot from the wall? 21. Who struck Abimelek, son of Jerubbaal? Didn't a woman throw an upper millstone on him from the wall so he died at Thebez? Why did you come up to the wall?' then you'll say, 'Also, your servant, Uriah, the Hittite, is dead.'"

22. And the messenger went; and he came and told David everything that Joab had sent him. 23. And the messenger said to David, "Because the people overpowered us and came out to us at the field, and we were on them up to the entrance of the gate, 24. and the archers shot at your servants from on the wall, and some of the king's servants died. And also, your servant, Uriah, the Hittite, is dead."

25. And David said to the messenger, "Say this to Joab: 'Don't let this thing be bad in your eyes, because the sword eats up one the same as another. Make your battle against the city stronger and tear it down.' And strengthen him."

26. And Uriah's wife heard that Uriah, her man, had died, and she mourned over her husband. 27. And the mourning passed, and David sent and collected her to his house, and she became his wife, and she gave birth to a son by him.

And the thing that David had done was bad in YHWH's eyes.

12:1. And YHWH sent Nathan to David, and he came to him and said to him, "There were two men in a city, one rich and one poor. 2. The rich one had very many sheep and oxen, 3. and the poor one didn't have anything except one little lamb, which he had bought and kept alive. And it had grown up with him and with his children together. It ate from his piece of bread and drank from his cup and lay in his bosom and was like a daughter to him. 4. And a traveler came to the rich man, and he held back from taking from his sheep and from his oxen to make for the guest who came to him, and he took the poor man's lamb and made it for the man who came to him."

5. And David's anger flared at the man very much, and he said to Nathan, "As YHWH lives, the man who did this is a dead man! 6. And he'll pay four times the lamb because he did this thing and didn't hold back."

7. And Nathan said to David, "You're the man! YHWH, God of Israel, said this: 'I anointed you to be king over Israel, and I rescued you from Saul's hand, 8. and I gave you your lord's house and your lord's wives in your bosom, and I gave you the house of Israel and Judah. And if this were too small, I would have added this much to you and this much again. 9. Why did you disdain YHWH's word, to do bad in my eyes? You struck Uriah, the Hittite, with the sword, and you took his wife for yourself as a wife, and you killed him with the sword of the children of Ammon. 10. And now: the sword won't ever turn away from your house, because you disdained me and took the wife of Uriah, the Hittite, to become yours as a wife.' 11. YHWH said this: 'Here, I'm setting up bad against you from your house. And I'll take your wives before your eyes and give them to your fellow man, and he'll lie with your wives in the eyes of this sun! 12. Because *you* did it in secret, but *I'll* do this thing in front of all Israel and in front of the sun.'"

13. And David said to Nathan, "I've sinned against YHWH."

And Nathan said to David, "Also, YHWH has caused your sin to pass. You won't die. 14. Nevertheless: because you *rejected* [the enemies of] YHWH in this thing, the son who is born to you will *die.*"15. And Nathan went to his house.

And YHWH struck the child to whom Uriah's wife had given birth by David, and he became ill. 16. And David asked God for the boy's sake, and David fasted and came in and spent the night and lay on the ground. 17. And the elders of his house stood over him to get him up from the ground, and he was not willing and would not eat bread with them. 18. And it was on the seventh day, and the child died. And David's servants were afraid to tell him that the child was dead, because they said, "Here, when the child was alive we spoke to him, and he didn't listen to our voice, and how shall we say to him, 'The child is dead,' and he'll do something bad!"

19. And David saw that his servants were whispering, and David understood that the child was dead. And David said to his servants, "Is the child dead?"

And they said, "He's dead."

20. And David got up from the ground and washed and anointed himself and changed his clothing and came to the house of YHWH and bowed.

And he came to his house and asked, and they set bread out for him, and he ate. 21. And his servants said to him, "What is this thing that you've done? You fasted and wept for the child when he was alive, and when the child was dead you got up and ate bread."

22. And he said, "As long as the child was alive I fasted and wept because I said, 'Who knows if YHWH will be gracious to me, and the child will live?' 23. And, now he's dead, why is this that I should fast? Am I able to bring him back again? *I'm* going to *him. He* won't come back to *me.*"

24. And David consoled Bathsheba, his wife, and came to her and lay with her, and she gave birth to a son. And he called his

name Solomon. And YHWH loved him. 25. And He sent by the hand of Nathan, the prophet. And he called his name Jedidiah because of YHWH.

26. And Joab fought against Rabbah of the children of Ammon and captured the royal city. 27. And Joab sent messengers to David and said, "I fought against Rabbah. I've also captured the city of the water. 28. And now, gather the rest of the people and camp against the city and capture it, rather than if I should capture the city, and it would be called by my name."

29. And David gathered all the people and went to Rabbah and fought against it and captured it. 30. And he took their king's crown from his head, and its weight was a talent of gold, with precious stones, and it was on David's head. And he brought out the city's spoil, very much. 31. And he brought out the people who were in it and put it to the saw and to sharp iron tools and to iron axes, and he passed them into the brick-mold. And he did this to all the cities of the children of Ammon. And David and all the people came back to Jerusalem.

13:1. And it was after this, and Absalom, son of David, had a beautiful sister, and her name was Tamar. And Amnon, son of David, loved her. 2. And Amnon was so distressed as to make himself sick over Tamar, his sister, because she was a virgin, and it was inconceivable in Amnon's eyes to do anything to her.

3. And Amnon had a friend, and his name was Jonadab, son of Shimeah, David's brother, and Jonadab was a very smart man. 4. And he said to him, "Why are you so low, son of the king, morning after morning? Won't you tell me?"

And Amnon said to him, "I love Tamar, my brother Absalom's sister."

5. And Jonadab said to him, "Lie down on your bed and act sick, and your father will come to see you, and say to him, 'Let Tamar, my sister, come and feed me bread, and she'll make the food before my eyes so I'll see, and I'll eat from her hand.'"

6. And Amnon lay down and acted sick, and the king came to see him, and Amnon said to the king, "Let Tamar, my sister, come and make two pancakes before my eyes, and let me eat from her hand."

7. And David sent to Tamar at the house, saying, "Go to your brother Amnon's house and make some food for him."

8. And Tamar went to her brother Amnon's house. And he was lying down. And she took the dough and kneaded it and made pancakes before his eyes and cooked the pancakes, 9. and she took the pan and dished it out in front of him. And he refused to eat. And Amnon said, "Have everyone go out from me." And everyone went out from him. 10. And Amnon said to Tamar, "Bring the food to the room, so I may eat from your hand." And Tamar took the pancakes that she had made and brought them to the room, to Amnon, her brother, 11. and she brought them over to him to eat.

And he took hold of her and said to her, "Come, lie with me, my sister!"

12. And she said to him, "Don't, my brother! Don't degrade me. Because such a thing isn't done in Israel. Don't do this foolhardy thing. 13. And I, where would I bear my disgrace? And *you'll* be like one of the fools in Israel. And now, speak to the king, because he won't hold me back from you." 14. And he was not willing to listen to her voice. And he overpowered her, and he degraded her and lay with her.

15. And Amnon felt a very big hatred for her, so that the hatred with which he hated her was bigger than the love with which he

had loved her. And Amnon said to her, "Get up. Go."

16. And she said to him, "Don't. This bad thing—to send me away—is bigger than the other thing that you did to me."

And he was not willing to listen to her. 17. And he called his boy who attended him and said, "Send this one away from me, outside! And lock the door behind her!"

18. And she had on a coat of many colors, because the king's virgin daughters would wear robes like that. And his attendant brought her outside and locked the door behind her. 19. And Tamar put ashes on her head and ripped the coat of many colors that she had on and put her hand to her head and went on and on screaming. 20. And Absalom, her brother, said to her, "Has Amnon, your brother, been with you? And now, my sister, keep quiet. He's your brother. Don't set your heart on this thing." And Tamar stayed at her brother Absalom's house and was devastated. 21. And King David heard all these things and was very furious. 22. And Absalom did not speak with Amnon, either bad or good, because Absalom hated Amnon over the fact that he had degraded Tamar, his sister.

23. And it was two years later, and they were shearing for Absalom in Baal-hazor, which is by Ephraim, and Absalom called all the king's sons. 24. And Absalom came to the king and said, "Here, they're shearing for your servant. Let the king and his servants go with your servant."

25. And the king said to Absalom, "No, my son, let's not all go, so we won't be a burden on you." And he pressed him, but he was not willing to go, and he blessed him.

26. And Absalom said, "And if not, let Amnon, my brother, go with us."

And the king said to him, "Why should he go with you?" 27. And Absalom pressed him, and he sent Amnon and all the king's sons with him.

28. And Absalom commanded his young men, saying, "See when Amnon's heart is in a good mood from wine, and I'll say to you, 'Strike Amnon!' And you'll kill him. Don't be afraid. Isn't it because *I've* commanded you? Be strong and be warriors." 29. And Absalom's young men did to Amnon as Absalom had commanded.

And all the king's sons got up and rode, each on his mule, and fled. 30. And they were on the way, and the news came to David, saying, "Absalom has struck all the king's sons, and not one of them is left." 31. And the king got up and ripped his clothes and lay down on the ground. And all his servants were standing with ripped clothes.

32. And Jonadab, son of Shimeah, David's brother, answered, and he said, "Let my lord not say, 'They killed all the young men, the king's sons,' because Amnon alone is dead, because this was set by Absalom since the day he degraded Tamar, his sister. 33. And now, let my lord the king not take this thing to heart, saying, 'All the king's sons are dead,' because Amnon alone is dead."

34. And Absalom fled. And the young man who was on watch raised his eyes and saw, and here were many people coming from the road behind him along the mountainside. 35. And Jonadab said to the king, "Here come the king's sons! It was as your servant spoke." 36. And it was as he finished speaking, and here came the king's sons, and they raised their voices and wept, and the king and all his servants wept as well, a very big weeping.

37. And Absalom fled and went to Talmai, son of Ammihud, king of Geshur. And King David mourned over his son all the time.

38. And Absalom fled and went to Geshur and was there three years. 39. And King David longed to go out to Absalom, because he was consoled over Amnon because he was dead. 14:1. And Joab,

son of Zeruiah, sensed that the king's heart was upon Absalom. 2. And Joab sent to Tekoa and took a wise woman from there and said to her, "Mourn, and wear mourning clothes, and don't anoint yourself with oil, and be like a woman who's been mourning many days now over the dead. 3. And you'll come to the king and speak to him like this." And Joab put the words in her mouth.

4. And the woman from Tekoa said it to the king: and she fell with her nose to the ground and bowed and said, "Save, O king!"

5. A the king said to her, "What business do you have?"

And she said, "But I'm a widow woman, and my husband died, 6. and your maid has two sons, and the two of them fought in the field, and there was no one with them to rescue, and one struck the other and killed him. 7. And, here, the whole family stood up against your maid and said, 'Give up the one who struck his brother, and we'll kill him for the life of his brother whom he killed, and let us kill the heir as well.' And they'll put out my burning coal that's left, so as not to provide a name and a remnant for my man on the face of the earth."

8. And the king said to the woman, "Go to your house, and I'll command regarding you."

9. And the woman from Tekoa said to the king, "My lord the king, let the crime be on me and on my father's house, and the king and his throne are innocent."

10. And the king said, "And whoever speaks to you, bring him to me, and he won't continue to touch you again."

11. And she said, "Let the king invoke YHWH, your God, that you won't let the blood avenger add to the annihilating, and they won't destroy my son."

And he said, "As YHWH lives, not a hair of your son's will fall to the ground."

12. And the woman said, "Let your maid speak a word to my lord the king."

And he said, "Speak."

13. And the woman said, "Why have you thought a thing like this about God's people, since from the king's speaking this thing *he* is guilty because of the king's not bringing back his banished. 14. Because we'll *die* and be like water that's spilled on the ground, that can't be gathered back, and God won't absolve a soul and have thoughts of not driving away a banished one from Him. 15. And now that I've come to speak this thing to the king, my lord, it's because the people made me afraid, so your maid said, 'Let me speak to the king. Maybe the king will do his maid's matter, 16. because the king will listen, to save his maid from the man's hand, to destroy me and my son together from God's legacy.' 17. And your maid said, 'Let the word of my lord the king, be a comfort, because my lord the king, is like an angel of God, to listen to the good and the bad.' And may YHWH, your God, be with you."

18. And the king answered, and he said to the woman, "Don't hide a thing from me that I ask you."

And the woman said, "Let my lord the king, speak."

19. And the king said, "Is Joab's hand with you in all of this?"

And the woman answered, and she said, "As your soul lives, my lord the king, no one can get to the right or the left of all that my lord the king has said, because your servant, Joab, he commanded me, and he put all these words in your maid's mouth. 20. Your servant, Joab, did this thing in order to turn the matter's appearance around. But my lord is wise, like the wisdom of an angel of God, to know everything that's in the land."

21. And the king said to Joab, "Here, I've done this thing: and go bring back the young man, Absalom!"

22. And Joab fell on his face to the ground and bowed and blessed the king, and Joab said, "Today your servant knows that I've found favor in your eyes, my lord the king, because the king

has performed your servant's word." 23. And Joab got up and went to Geshur and brought Absalom to Jerusalem.

24. And the king said, "Let him turn to his house and not see my face." And Absalom turned to his house and did not see the king's face.

25. And there was not a man as handsome as Absalom in all Israel—to praise very much. From his foot to the top of his head, there was not a blemish on him. 26. And when he cut the hair on his head (and it was at the end of regular periods that he cut it, because it was heavy on him, and he cut it), he weighed the hair from his head: two hundred shekels by the king's weight. 27. And three sons were born to Absalom and one daughter, and her name was Tamar. She was a beautiful woman.

28. And Absalom lived in Jerusalem two years and did not see the king's face. 29. And Absalom sent to Joab to send him to the king, and he was not willing to come to him. And he sent again, a second time, and he was not willing to come. 30. And he said to his servants, "See, Joab's property is near me, and he has barley there. Go and burn it with fire." And Absalom's servants burned the property with fire.

31. And Joab got up and came to Absalom in the house and said to him, "Why did your servants burn the property that I have with fire?"

32. And Absalom said to Joab, "Here, I sent to you, saying, 'Come here, and let me send you to the king, saying, "Why did I come from Geshur? It was good for me as long as I was there. And now, let me see the king's face, and if there's a crime in me then kill me!"'" 33. And Joab came to the king and told him. And he called Absalom, and he came to the king and bowed to him, on his nose, to the ground before the king. And the king kissed Absalom.

15:1. And it was after this, and Absalom made a chariot and horses for himself and fifty men running in front of him. 2. And Absalom got up early and stood by the road to the gate; and, whenever a man who had case would come to the king for judgment, Absalom would call to him and say, "From what city are you?"

And he would say, "Your servant is from one of the tribes of Israel."

3. And Absalom would say to him, "See, your words are good and right, and you don't have someone from the king to listen!" 4. And Absalom would say, "Who would make me judge in the land—and everyone who had a case or judgment could come to me, and I'd give him justice." 5. And it was, when a man would come close to bow to him, that he would put out his hand and take hold of him and kiss him. 6. And Absalom acted in this way to all Israel who would come for judgment to the king, and Absalom stole the heart of the people of Israel.

7. And it was at the end of four years, and Absalom said to the king, "Let me go, and I'll pay my vow that I made to YHWH in Hebron, 8. because your servant made a vow when I was living in Geshur in Aram, saying, 'If YHWH will *bring me back* to Jerusalem, then I'll serve YHWH.'"

9. And the king said to him, "Go in peace." And he got up and went to Hebron.

10. And Absalom sent spies in all the tribes of Israel, saying, "When you hear the sound of the ram's horn, then you shall say, 'Absalom rules in Hebron!'" 11. And two hundred men went with Absalom from Jerusalem, summoned, and they were going in innocence, and they did not know anything. 12. And Absalom sent and

summoned Ahitophel, the Gilonite, David's advisor, from his city, from Giloh, when he offered the sacrifices. And the conspiracy was powerful, and the people with Absalom were getting greater.

13. And someone came to tell David, saying, "The heart of the people of Israel is behind Absalom."

14. And David said to all his servants who were with him in Jerusalem, "Get up, and let's flee, because we won't have a survivor because of Absalom. Hurry to go or else he'll hurry and overtake us and thrust something bad on us and strike the city with the edge of the sword."

15. And the king's servants said to the king, "According to everything that my lord the king has chosen, your servants are here." 16. And the king went out with all of his house at his feet.

And the king left ten concubine women to watch over the house.

17. And the king went out with all the people at his feet, and they stopped at Beth-merhak. 18. And all his servants were passing alongside him, and all the Cherethites and all the Pelethites; and all the Gittites, six hundred men who came at his feet from Gath, were passing before the king. 19. And the king said to Ittai, the Gittite, "Why should *you* go with us as well? Go back and live with the king, because you're a foreigner, and you're an exile from your place as well. 20. You came yesterday, so should I make you roam today, to go with us as I'm going wherever I'm going?! Go back, and take your brothers back with you in kindness and faithfulness."

21. And Ittai answered the king, and he said, "As YHWH lives and as my lord the king lives: in the place where my lord the king will be, whether for death or for life, your servant will be there."

22. And David said to Ittai, "Go and pass." And Ittai, the Gittite, and all his people and all the infants who were with him, passed.

23. And all the land were weeping, a big voice, and all the people were passing, and the king was passing through the wadi Kidron,

and all the people were passing on the road through the wilderness. 24. And here was Zadok as well, and all the Levites with him, carrying the ark of God's covenant, and they set God's ark up, and Abiathar went up, until all the people had finished passing from the city. 25. And the king said to Zadok, "Take God's ark back to the city. If I find favor in YHWH's eyes, then He'll bring me back and show it and its place of residence to me. 26. And if He says this: 'I don't desire you,' here I am, let Him do to me whatever is good in His eyes." 27. And the king said to Zadok, the priest, "Do you see? Go back to the city in peace, and Ahimaaz, your son, and Jonathan, son of Abiathar, your two sons, with you. 28. See, I'm delaying in the plains of the wilderness until word comes from you to tell me." 29. And Zadok and Abiathar took God's ark back to Jerusalem, and they stayed there. 30. And David had gone up by the ascent of the Mount of Olives, going up and weeping, and his head was covered, and he was going barefoot, and all the people who were with him each had his head covered, and they went up, going up and weeping. 31. And David had been told, saying, "Ahitophel is among the conspirators with Absalom." And David said, "Make Ahitophel's advice foolish, YHWH." 32. And David was coming to the top, so that he could bow to God there, and here was Hushai, the Archite, coming to him, his coat torn and dirt on his head. 33. And David said to him, "If you pass with me, then you'll become a burden on me; 34. but if you go back to the city and say to Absalom, 'I'm your servant, King. I was your father's servant in the past, and now I'm your servant,' then you'll nullify Ahitophel's advice for me. 35. And aren't Zadok and Abiathar, the priests, with you there? And it will be that every word that you hear from the king's house you'll tell to Zadok and to Abiathar, the priests. 36. Here, their two sons, Ahimaaz of Zadok and Jonathan of Abiathar, are there with them, and you'll send every word that you hear to me by their hand." 37. And Hushai, David's companion, came to the city. And Absalom was coming to Jerusalem.

16:1. And David had passed a little from the top, and here was Ziba, Mephibaal's young man, coming to him, and a pair of saddled asses, and two hundred loaves of bread and a hundred raisin clusters and a hundred summer fruits and a bottle of wine. 2. And the king said to Ziba, "What are these that you have?"

And Ziba said, "The asses are for the king's house to ride, and the bread and the summer fruits are for the young men to eat, and the wine is for the fatigued to drink in the wilderness."

3. And the king said, "And where is your lord's son?"

And Ziba said to the king, "Here, he's staying in Jerusalem, because he said, 'Today the house of Israel will give me back my father's kingdom.'"

4. And the king said to Ziba, "Here, everything of Mephibaal's is yours."

And Ziba said, "I bow. May I find favor in your eyes, my lord the king."

5. And King David came to Bahurim, and here was a man from Saul's family coming out from there, and his name was Shimei, son of Gera; and he was coming out and cursing continually 6. and throwing stones at David and at all of King David's servants and all the people and all the warriors on his right and on his left. 7. And, in his cursing, Shimei would say this: "Out, out, man of blood! Good-for-nothing man! 8. YHWH has brought back on you all the blood of the house of Saul, in whose place you've ruled, and YHWH has delivered the kingdom into the hand of Absalom, your son! And here you are in your bad condition because you're a man of blood!"

9. And Abishai, son of Zeruiah, said to the king, "Why should this dead dog curse my lord the king? Let me cross over and remove his head!"

10. And the king said, "What do I have to do with you sons of Zeruiah? Let him curse like this. Because YHWH said to him, 'Curse David,' so who can say, 'Why have you done this?'" 11. And David said to Abishai and to all his servants, "Here, my son—who came out of my insides!—seeks my life. How much more this Benjaminite now! Leave him alone, and let him curse, because YHWH said it to him. 12. Maybe YHWH will look on my degradation, and YHWH will pay me back with good for his cursing this day." 13. And David and his people went on the road, and Shimei was going on the ridge of the mountain across from him, going on and on cursing, and he threw stones across at him and threw dirt.

14. And the king and all the people who were with him came, exhausted, and he was getting refreshed there.

15. And Absalom and all the people, the men of Israel, came to Jerusalem, and Ahitophel with him. 16. And it was when Hushai, the Archite, David's companion, came to Absalom: and Hushai said to Absalom, "Long live the king! Long live the king!"

17. And Absalom said to Hushai, "This is your kindness to your companion?! Why didn't you go with your companion?"

18. And Hushai said to Absalom, "No, but the one whom YHWH and this people and all the men of Israel have chosen, I shall be his, and I'll stay with him. 19. And, second, whom should I serve? Shouldn't it be before his son? As I served before your father, so I'll be before you."

20. And Absalom said to Ahitophel, "Give advice. What shall we do?"

21. And Ahitophel said to Absalom, "Come to your father's concubines whom he left to watch the house, and all Israel will hear that you've made yourself odious to your father, and the hands of

everyone who is with you will be strengthened." 22. And they pitched a tent for Absalom on the roof, and Absalom came to his father's concubines before all Israel's eyes.

23. And Ahitophel's advice that he gave in those days was like when a man asks the word of God. All of Ahitophel's advice was like this, both to David and to Absalom. 17:1. And Ahitophel said to Absalom, "Let me choose twelve thousand men and get up and pursue David tonight, 2. and I'll come upon him, and he'll be tired and his hands weak, and I'll make him terrified, and all the people who are with him will flee, and I'll strike the king by himself. 3. And let me bring all the people back to you. When all have come back—but the man whom you seek—all the people will be at peace."

4. And the thing was right in Absalom's eyes and in the eyes of all of Israel's elders. 5. And Absalom said, "Call Hushai, the Archite, as well, and let's hear what's in his mouth, him as well." 6. And Hushai, the Archite, came to Absalom, and Absalom said to him, saying, "Ahitophel spoke such a thing. Shall we do what he said? If not, you speak."

7. And Hushai said to Absalom, "The advice that Ahitophel has given isn't good this time." 8. And Hushai said, "You know your father and his men, that they're warriors, and they're bitter-souled, like a bear in the field bereaved of its children, and your father is a man of war, and he won't spend the night with the people. 9. Here, now he's hidden in one of the pits or in one of the places. And it will be that, the first time some of them fall and someone hears about it, he'll say, 'There's been a strike against the people who are behind Absalom.' 10. And even if he's a powerful man who has a heart like a lion's heart, he'll *melt,* because all Israel knows that your father is a warrior and those who are with him are powerful men. 11. Rather, I advise that all Israel be *gathered* to you, from Dan to Beer-sheba, like the sand that's by the sea for its great number, and

you going personally in the battle. 12. And we'll come to him in one of the places in which he's found, and we'll be on him like the dew falls on the ground, and not one will be left of him and of all the people who are with him. 13. And if he'll be gathered into a city, then all Israel will overtake that city with ropes, and we'll drag it to a wadi, so that not even a pebble will be found there."

14. And Absalom and all the people of Israel said, "The advice of Hushai, the Archite, is better than the advice of Ahitophel."

And YHWH had arranged to nullify Ahitophel's good advice in order for YHWH to bring a bad thing to Absalom.

15. And Hushai said to Zadok and to Abiathar, the priests, "Ahitophel advised Absalom and Israel's elders such and such, and I advised such and such. 16. And now, send quickly and tell David, saying, 'Don't spend the night in the plains of the wilderness, but *cross over,* or else the king and all the people who are with him will be swallowed up.'"

17. And Jonathan and Ahimaaz had been standing at En-rogel, and a maid went and told them, and they went and told King David—because they could not be seen coming to the city. 18. And a boy saw them, and he told Absalom. And the two of them went quickly and came to the house of a man in Bahurim; and he had a well in his yard, and they went down there. 19. And a woman took and spread a covering on the top of the well and spread out fruits on it, and nothing was known. 20. And Absalom's servants came to the woman at the house and said, "Where are Ahimaaz and Jonathan?"

And the woman said to them, "They crossed the water stream."

And they looked and did not find them, and they went back to Jerusalem. 21. And it was after they went, and they came up from the well and went and told King David. And they said to David, "Get up and cross the water quickly, because Ahitophel advised

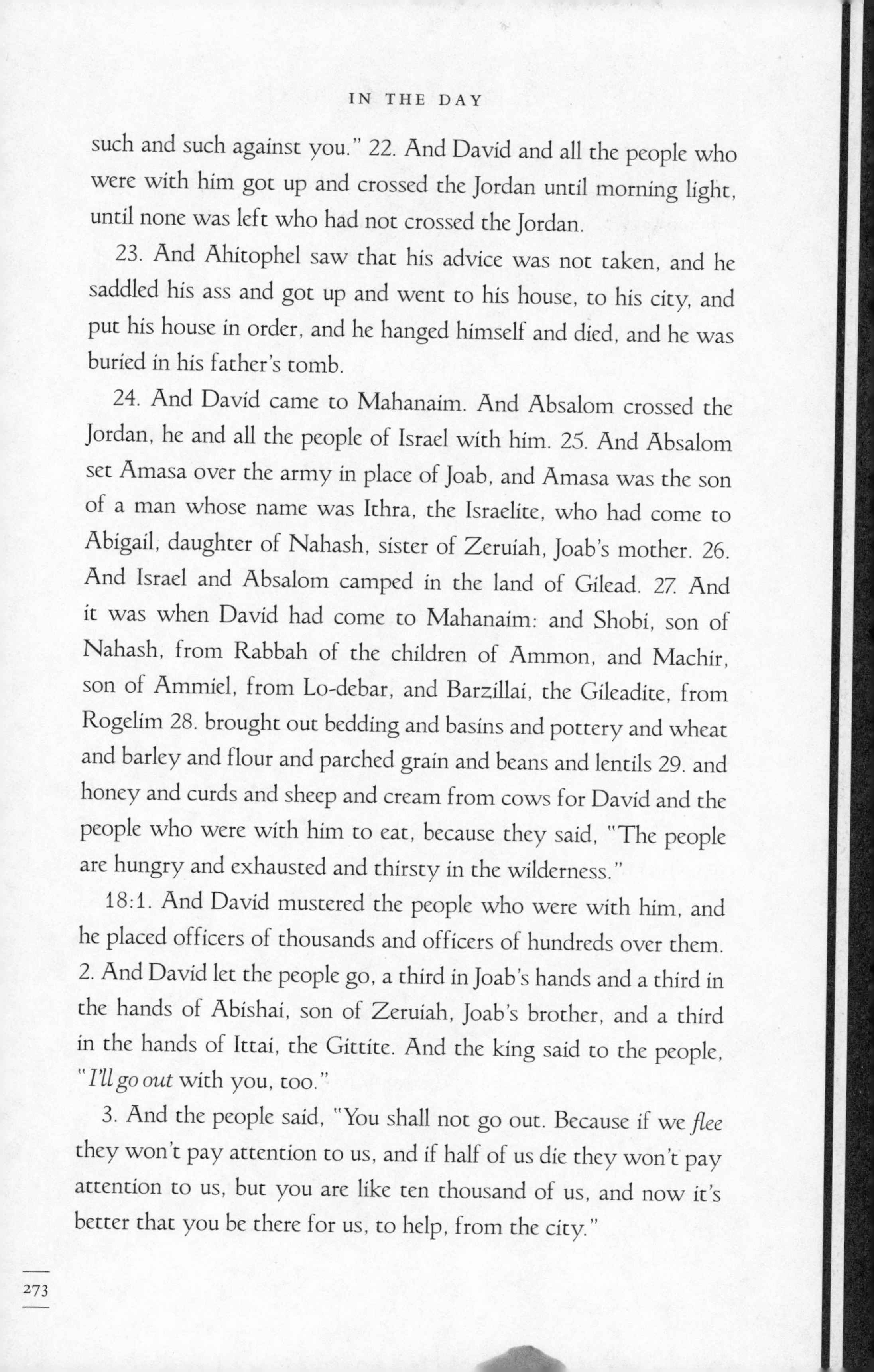

such and such against you." 22. And David and all the people who were with him got up and crossed the Jordan until morning light, until none was left who had not crossed the Jordan.

23. And Ahitophel saw that his advice was not taken, and he saddled his ass and got up and went to his house, to his city, and put his house in order, and he hanged himself and died, and he was buried in his father's tomb.

24. And David came to Mahanaim. And Absalom crossed the Jordan, he and all the people of Israel with him. 25. And Absalom set Amasa over the army in place of Joab, and Amasa was the son of a man whose name was Ithra, the Israelite, who had come to Abigail, daughter of Nahash, sister of Zeruiah, Joab's mother. 26. And Israel and Absalom camped in the land of Gilead. 27. And it was when David had come to Mahanaim: and Shobi, son of Nahash, from Rabbah of the children of Ammon, and Machir, son of Ammiel, from Lo-debar, and Barzillai, the Gileadite, from Rogelim 28. brought out bedding and basins and pottery and wheat and barley and flour and parched grain and beans and lentils 29. and honey and curds and sheep and cream from cows for David and the people who were with him to eat, because they said, "The people are hungry and exhausted and thirsty in the wilderness."

18:1. And David mustered the people who were with him, and he placed officers of thousands and officers of hundreds over them. 2. And David let the people go, a third in Joab's hands and a third in the hands of Abishai, son of Zeruiah, Joab's brother, and a third in the hands of Ittai, the Gittite. And the king said to the people, "*I'll go out* with you, too."

3. And the people said, "You shall not go out. Because if we *flee* they won't pay attention to us, and if half of us die they won't pay attention to us, but you are like ten thousand of us, and now it's better that you be there for us, to help, from the city."

4. And the king said to them, "I'll do whatever is good in your eyes." And the king stood by the gate, and all the people went out by hundreds and by thousands. 5. And the king commanded Joab and Abishai and Ittai, saying, "Go easy for me on the boy, on Absalom." And all the people heard when the king commanded all the officers on the matter of Absalom. 6. And the people went out to the field toward Israel, and the war was in the forest of Ephraim. 7. And the people of Israel were stricken there before David's servants, and the strike there was big in that day: twenty thousand. 8. And the war there was scattered over all the land, and the forest consumed a greater number of the people than the sword consumed in that day.

9. And Absalom happened to be in front of David's servants, and Absalom was riding on a mule, and the mule came under a branch of a big oak, and his head got caught in the oak, and he was suspended between the skies and the earth, and the mule that was under him passed on. 10. And a man saw it and told Joab, and he said, "Here, I saw Absalom hanging in an oak!"

11. And Joab said to the man who told him, "And here, you saw it, and why didn't you strike him to the ground there—and I would have had to give you ten weights of silver and a sash!"

12. And the man said to Joab, "And even if I could have a thousand weights of silver on my hands I wouldn't put out my hand at the king's son, because the king commanded you and Abishai and Ittai in our ears, saying, 'Watch out, whoever is against the boy, Absalom.' 13. Or, if I had done a dishonest thing against my life, since nothing is hidden from the king, then you would have stood by."

14. And Joab said, "Let me not wait around in front of you like this!" And he took three shafts in his hand and drove them into Absalom's heart. He was still alive in the heart of the oak, 15. and ten young men, Joab's armor-bearers, went around and struck

Absalom and killed him. 16. And Joab blasted on a ram's horn, and the people came back from pursuing Israel because Joab held the people back. 17. And they took Absalom and threw him into a big pit in the forest and set up a very big pile of stones over him. And all Israel had fled, each to his tent. 18. And Absalom had taken a pillar and set it up for himself while he was alive, which is in the valley of the king, "because," he said, "I don't have a son to make my name remembered." And he called the pillar by his name, and it has been called Absalom's Monument to this day.

19. And Ahimaaz, son of Zadok, said, "Let me run and give the king the news that YHWH has vindicated him from his enemies' hand."

20. And Joab said to him, "You are not a man giving news on this day. You'll give news on some other day, and on this day you won't give news because of the fact that the king's son is dead." 21. And Joab said to a Cushite, "Go tell the king what you've seen." And the Cushite bowed to Joab and ran.

22. And Ahimaaz, son of Zadok, went on again and said to Joab, "And whatever will be, let me run after the Cushite, I, too."

And Joab said, "What will you run for, my son, since you have no news to be found?"

23. "And whatever will be, I'll run."

And he said to him, "Run."

And Ahimaaz ran by way of the plain, and he passed the Cushite. 24. And David was sitting between the two gates, and the lookout went to the roof of the gate, to the wall, and raised his eyes and saw, and here was a man running alone. 25. And the lookout called and told the king. And the king said, "If he's alone, there's news in his mouth." And he went, coming closer.

26. And the lookout saw another man running, and the lookout called to the gatekeeper and said, "Here's a man running alone."

And the king said, "This one has news, too."

27. And the lookout said, "I recognize the running of the first one as the running of Ahimaaz, son of Zadok."

And the king said, "This is a good man, and he'll come with good news."

28. And Ahimaaz called, and he said to the king, "Peace," and he bowed to the king with his nose to the ground. And he said, "Blessed is YHWH, your God, who has closed in on the people who raised their hand against my lord the king!"

29. And the king said, "Is the boy, Absalom, all right?"

And Ahimaaz said, "I saw a big clamor when the king's servant, Joab, was sending your servant, and I didn't know what it was."

30. And the king said, "Turn around. Stand here." And he turned around and stood.

31. And here was the Cushite coming, and the Cushite said, "Let my lord the king receive the good news that YHWH has vindicated you today from the hand of all those who rose against you!"

32. And the king said to the Cushite, "Is the boy, Absalom, all right?"

And the Cushite said, "May my lord the king's enemies and all who rose against you for bad be like the boy."

19:1. And the king was distressed, and he went on the place over the gate and wept. And he said this as he went: "My son, Absalom, my son, my son, Absalom! Who could make it *my* dying, me in place of you, Absalom, my son, my son!"

2. And it was told to Joab: "Here, the king is weeping and mourning over Absalom." 3. And the salvation on that day became mourning for all the people because the people heard on that day, saying, "The king is grieved over his son." 4. And the people stole their way into the city on that day the way people who are ashamed steal away when they flee from battle.

5. And the king covered his face, and the king cried in a loud voice, "My son, Absalom, Absalom, my son, my son!"

6. And Joab came to the king at the house and said, "Today you've shamed the faces of all your servants who saved your life today and your sons' and daughters' lives and your wives' lives and your concubines' lives 7. by loving those who hated you—and hating those who love you!—because today you've conveyed that you don't have officers and servants, because today I know that, if Absalom were alive and we were all dead today, that would be right in your eyes! 8. And now, get up! Go out! And speak on your servants' hearts, because I swear to YHWH that, if you don't go out, no one will spend this night with you, and this will be worse for you than all the bad things that have come upon you from your childhood until now."

9. And the king got up and sat at the gate. And they told all the people, saying, "Here's the king, sitting at the gate!" And all the people came before the king.

And Israel had fled, each to his tent. 10. And the whole people was in contention in all the tribes of Israel, saying, "The king had rescued us from our enemy's hand, and he had freed us from the Philistines' hand, and now he has fled from the land, from Absalom. 11. And Absalom, whom we anointed over us, has died in the war. And now, why are you keeping quiet about bringing back the king?"

12. And King David sent to Zadok and to Abiathar, the priests, saying, "Speak to Judah's elders, saying, 'Why should you be last to bring back the king to his house? And the word of all of Israel has come to the king, to his house. 13. You're my brothers. You're my bone and my flesh. And why should you be last to bring back the king?' 14. And say to Amasa, 'Aren't you my bone and my flesh. May God do this to me and double this if you won't be commander of the army before me for all time in place of Joab.'"

15. And he turned the heart of all the people of Judah like one man, and they sent to the king: "Come back, you and all your servants." 16. And the king came back. And he came to the Jordan, and Judah had come to Gilgal to go to the king, to bring the king across the Jordan. 17. And Shimei, son of Gera, the Benjaminite, who was from Bahurim, hurried and came down with the people of Judah to King David, 18. and a thousand men with him from Benjamin; and Ziba, the steward of the house of Saul, and his fifteen sons and his twenty servants with him. And they rushed to the Jordan before the king, 19. and they crossed the ford to bring the king's household across and to do whatever was good in his eyes. And Shimei, son of Gera, fell in front of the king when he crossed the Jordan, 20. and said to the king, "Let my lord not impute a crime to me, and don't remember what your servant committed on the day that my lord the king went out of Jerusalem, that the king should pay attention to it, 21. because your servant knows that I sinned, and I've come here today first out of all the house of Joseph to come down to my lord the king."

22. And Abishai, son of Zeruiah, answered, and he said, "Will Shimei not be put to death for this: that he cursed YHWH's anointed?"

23. And David said, "What do I have to do with you sons of Zeruiah, that you should become an adversary to me today? Should any man in Israel be put to death today? Because don't I know that today I'm king over Israel?" 24. And the king said to Shimei, "You won't die!" And the king swore it to him.

25. And Mephibaal, son of Saul, had come down to the king. And he had not dressed his legs and had not trimmed his mustache and had not washed his clothes from the day the king had gone to the day that he came in peace. 26. And it was when he came to Jerusalem to the king: and the king said to him, "Why didn't you go with me, Mephibaal?"

27. And he said, "My lord the king, my servant deceived me, because your servant said, 'Let me saddle me an ass, and I'll ride on it and go with the king,' because your servant is a cripple. 28. And he betrayed your servant to my lord the king. And my lord the king is like an angel of God. And do whatever is good in your eyes. 29. Because my father's whole house was nothing but people condemned to death to my lord the king, and you set your servant among those who eat at your table. And what further merit and further grounds to cry out to the king do I have?"

30. And the king said to him, "Why do you speak your words further? I've said: You and Ziba will split the property."

31. And Mephibaal said to the king, "Let him even take it all, since my lord the king has come in peace to his house."

32. And Barzillai, the Gileadite, came down from Rogelim and passed with the king to the Jordan to send him off him at the Jordan. 33. And Barzillai was very old—eighty years old—and he had provided for the king when he was staying at Mahanaim, because he was a very big man. 34. And the king said to Barzillai, "You cross with me, and I'll provide for you with me in Jerusalem."

35. And Barzillai said to the king, "How many are the days of the years of my life that I should go up with the king to Jerusalem? 36. I'm eighty years old today. Will I know the difference between good and bad? Will your servant taste what I eat and what I drink? Will I still hear the sound of men singing? and women singing? So why should your servant become more of a burden to my lord the king? 37. Your servant will just cross the Jordan with the king, and why should the king give me this compensation? 38. Let your servant go back, and I'll die in my city, by my father's and my mother's tomb. And here's your servant, Chimham. Let him cross with my lord the king, and do for him whatever is good in your eyes."

39. And the king said, "Chimham will cross with me, and I'll do for him whatever is good in *your* eyes, and I'll do for you whatever you choose from me."

40. And all the people crossed the Jordan, and the king crossed, and the king kissed Barzillai and blessed him, and he went back to his place. 41. And the king passed to Gilgal, and Chimham passed with him, and all the people of Judah escorted the king, and also half of the people of Israel. 42. And here were all the people of Israel coming to the king, and they said to the king, "Why did our brothers, the people of Judah, steal you and bring the king and his house across the Jordan, and all of David's people with him?"

43. And all the people of Judah answered the people of Israel, "Because the king is close to me. And why is this that you're angry over this thing? Have we *eaten* anything of the king's? Did he *bear* anything for us?"

44. And the people of Israel answered the people of Judah, and they said, "I have ten shares in the king, and I have more in David than you, and why did you take me lightly, that I wouldn't have my word first to bring back my king?"

And the word of the people of Judah was harsher than the word of the people of Israel.

20:1. And a good-for-nothing man was called there, and his name was Sheba, son of Bichri, a Benjaminite, and he blasted on a ram's horn and said, "We don't have any portion in David, and we don't have a legacy in the son of Jesse! Each man to his tents, Israel!" 2. And all the people of Israel went up from being behind David to being behind Sheba, son of Bichri; but the people of Judah had clung to their king, from the Jordan to Jerusalem.

3. And David came to his house at Jerusalem, and the king took the ten concubine women whom he had left to watch the house,

and he put them in a watch-house. And he provided for them. And he did not come to them, and they were restricted until the day of their death, a living widowhood.

4. And the king said to Amasa, "Call the men of Judah to me in three days, and you stay here." 5. And Amasa went to call Judah, and he took longer than the time that he had appointed for him. 6. And David said to Abishai, "Now Sheba, son of Bichri, will do worse to us than Absalom. You take your lord's servants and pursue him, or else he'll find fortified cities for himself and snatch away our eye." 7. And Joab's men and the Cheretites and the Pelethites and all the warriors went out behind him, and they came out of Jerusalem to pursue Sheba, son of Bichri.

8. They had been at the big stone that is in Gibeon when Amasa came in front of them. And Joab was wearing his outfit, and he was wearing a sword fastened on his hip in a sheath, and when he had come out it slid down. 9. And Joab said to Amasa, "Are you all right, my brother?" And Joab's right hand took hold of Amasa's beard to kiss him. 10. And Amasa was not watchful of the sword that was in Joab's hand. And he struck him to the abdomen, and his insides spilled to the ground, and he did not strike him a second time, and he died. And Joab and Abishai, his brother, pursued Sheba, son of Bichri. 11. And as a man from among Joab's young men stood over him he said, "Whoever desires Joab and whoever is for David: After Joab!" 12. And the man saw that while Amasa was wallowing in blood in the highway all the people stood still. And the man took Amasa out of the highway to a field and threw clothing over him since he saw that everyone who came upon him stood still. 13. When he had been cast from the highway, everyone passed on after Joab to pursue Sheba, son of Bichri.

14. And he passed through all the tribes of Israel to Abel and Beth-maacah and all the Berites, and they assembled and came after him as well. 15. And they came and besieged him in Abel of Beth-maacah

and cast up a mound against the city, and it stood in a fortification. And all the people who were with Joab were destroying, to knock down the wall.

16. And a wise woman called from the city, "Listen! Listen! Say to Joab, 'Come close to here, and let me speak to you.'" 17. And he came close to her. And the woman said, "Are you Joab?"

And he said, "I am."

And she said to him, "Listen to your maid's words."

And he said, "I'm listening."

18. And she said, saying, "In olden days they would *speak*, saying, 'They should *ask* in Abel!' and they would bring a matter to a close this way. 19. I'm among Israel's peaceful, faithful ones. You're seeking to put a city and a mother in Israel to death. Why should you swallow YHWH's legacy?"

20. And Joab answered, and he said, "Far be it! Far be it from me that I should swallow or I should destroy. 21. The matter is not like that. But a man from the mountains of Ephraim—Sheba, son of Bichri, is his name—raised his hand against King David. Surrender him alone, and let me go from against the city."

And the woman said to Joab, "Here, his head will be thrown to you from the wall." 22. And the woman came to all the people with her wisdom, and they cut off the head of Sheba, son of Bichri, and threw it to Joab. And he blasted on the ram's horn, and they scattered from the city, each to his tents, and Joab went back to Jerusalem, to the king.

23. And Joab was over all the army of Israel; and Benaiah, son of Jehoiada, was over the Cherethites and the Pelethites.

1 Kings 1:1. And King David was old, well along in days. And they covered him in clothes, and he did not get warm. 2. And his

servants said to him, "Let them seek a virgin young woman for my lord the king, and she'll stand before the king and be a help to him, and she'll lie in your bosom, and my lord the king will get warm." 3. And they sought a beautiful young woman in all the borders of Israel, and they found Abishag, the Shunammite, and they brought her to the king. 4. And the young woman was extremely beautiful, and she was a help to the king, and she served him. And the king did not know her.

5. And Adonijah, son of Haggith, promoted himself, saying, "I'll rule." And he made a chariot and horsemen for himself, and fifty men running ahead of him. 6. And his father had never made him suffer in his life, saying, "Why did you act like that?" And he was very good-looking as well. And she gave birth to him after Absalom. 7. And his words were with Joab, son of Zeruiah, and with Abiathar, the priest; and they supported Adonijah. 8. And Zadok, the priest, and Benaiah, son of Jehoiada, and Nathan, the prophet, and Shimei and Rei and the warriors whom David had were not with Adonijah. 9. And Adonijah sacrificed sheep and oxen and fatlings at the stone of Zoheleth, which is near En-rogel. And he called all his brothers, the king's sons, and all the people of Judah, the king's servants, 10. and he did not call Nathan, the prophet, and Benaiah and the warriors and Solomon, his brother.

11. And Nathan said to Bathsheba, Solomon's mother, saying, "Haven't you heard that Adonijah, son of Haggith, reigns, and our lord David doesn't know? 12. And now, go. Let me give you advice, and you'll save your life and the life of your son, Solomon. 13. Go. And come to King David and say to him, 'Didn't you, my lord the king, swear to your maid, saying, "Solomon, your son, will rule after me, and he'll sit on my throne"? So why does Adonijah reign?' 14. Here, while you're still speaking there with the king, I'll come after you and confirm your words."

15. And Bathsheba came to the king in the room. And the king was very old. And Abishag, the Shunammite, was serving the king. 16. And Bathsheba knelt and bowed to the king. And the king said, "What do you want?"

17. And she said to him, "My lord, you swore by YHWH, your God, to your maid, 'Solomon, your son, will reign after me, and he'll sit on my throne.' 18. And now, here, Adonijah reigns; and now, my lord the king, you didn't know! 19. And he has sacrificed a great number of oxen and fatlings and sheep, and he has called all the king's sons and Abiathar, the priest, and Joab, the commander of the army, and he didn't call Solomon, your servant. 20. And you, my lord the king: the eyes of all of Israel are on you to tell them who will sit on my lord the king's throne after him. 21. And it will be, when my lord the king will be lying with his fathers, that I and my son, Solomon, will be criminals!"

22. And, here, she was still speaking with the king, when Nathan, the prophet, came. 23. And they told the king, saying, "Here's Nathan, the prophet." And he came before the king and bowed to the king with his nose to the ground. 24. And Nathan said, "My lord the king, did you say, 'Adonijah will reign after me, and he'll sit on my throne'? 25. Because he went down today and sacrificed a great number of oxen and fatlings and sheep, and he called all the king's sons and the officers of the army and Abiathar, the priest. And here they are, eating and drinking in front of him, and they've said, 'Long live King Adonijah.' 26. And he didn't call me—me, your servant—and Zadok, the priest, and Benaiah, son of Jehoiada, and Solomon, your servant. 27. Has this thing come about from my lord the king, and you haven't informed your servant who would sit on my lord the king's throne after him?"

28. And King David answered, and he said, "Call Bathsheba to me."

And she came before the king, and she stood before the king. 29. And the king swore, and he said, "As YHWH lives, who redeemed my life from every trouble: 30. as I swore to you by YHWH, God of Israel, saying, 'Solomon, your son, will reign after me, and he'll sit on my throne in my place,' so I shall do this day!"

31. And Bathsheba knelt, nose to the ground, and bowed to the king and said, "Long live my lord King David forever!"

32. And King David said, "Call Zadok, the priest, and Nathan, the prophet, and Benaiah, son of Jehoiada, to me."

And they came before the king. 33. And the king said to them, "Take your lord's servants with you, and have Solomon, my son, ride on a mule that is mine and bring him down to Gihon. 34. And let Zadok, the priest, and Nathan, the prophet anoint him there as king over Israel, and you shall blast on a ram's horn and say, 'Long live King Solomon!' 35. And you shall go up behind him, and he'll come and sit on my throne, and he'll reign in my place. And I've commanded him to be the designated king over Israel and over Judah."

36. And Benaiah, son of Jehoiada, answered the king, and he said, "Amen! So may YHWH, my lord the king's God, say! 37. As YHWH was with my lord the king, so may He be with Solomon! And may He make his throne greater than my lord King David's throne!"

38. And Zadok, the priest, and Nathan, the prophet, and Benaiah, son of Jehoiada, and the Cherethites and the Pelethites went down and had Solomon ride on King David's mule and brought him to Gihon. 39. And Zadok, the priest, took a horn of oil from the tent and anointed Solomon. And they blasted on a ram's horn, and all the people said, "Long live King Solomon!" 40. And all the people went up behind him, and the people were playing flutes and were happy, a big happiness. And the earth was broken up at the sound of them.

41. And Adonijah and everyone who had been called who were with him heard it. And they had finished eating, and Joab heard the sound of the ram's horn and said, "Why is the sound of the city stirred up?"

42. He was still speaking, and here was Jonathan, son of Abiathar, the priest, coming. And Adonijah said, "Come, because you're a valiant man, and you bring good news."

43. And Jonathan answered, and he said to Adonijah, "But our lord King David has made Solomon king! 44. And the king has sent Zadok, the priest, and Nathan, the prophet, and Benaiah, son of Jehoiada, and the Cherethites and the Pelethites, and they had him ride on the king's mule. 45. And Zadok, the priest, and Nathan, the prophet, have anointed him as king at Gihon, and they've gone up from there happily, and the city is tumultuous, and that's the sound that you heard. 46. And, also, Solomon has sat on the royal throne. 47. And, also, the king's servants have come to bless our lord King David, saying, 'May God make Solomon's name better than yours, and may He make his throne greater than yours!' And the king bowed on the bed. 48. And, also, the king said this: 'Blessed is YHWH, God of Israel, who has provided someone to sit on my throne today, and my eyes see it!'"

49. And they trembled. And everyone who had been called by Adonijah got up and went, each his own way. 50. And Adonijah was afraid of Solomon, and he got up and went and took hold of the horns of the altar.

51. And it was told to Solomon, saying, "Here, Adonijah fears King Solomon; and, here, he has taken hold of the horns of the altar, saying, 'Let King Solomon swear to me this day that he won't kill his servant by the sword!'"

52. And Solomon said, "If he'll be a valiant man, not a hair of his will fall to the earth; and if some bad will be found in him, then he'll

die." 53. And King Solomon sent, and they brought him down from the altar. And he came and bowed to King Solomon. And Solomon said to him, "Go to your house."

2:1. And David's time to die drew close. And he commanded Solomon, his son, saying, 2. "I'm going in the way of all the earth. And you shall be strong and become a man. 5. And, also, you know what Joab, son of Zeruiah, did to me, what he did to the two commanders of Israel's army, to Abner, son of Ner, and to Amasa, son of Jether: that he killed them, and he shed the blood of war in peace, and he put the blood of war in his belt that was on his hips and in his shoe that was on his feet. 6. And you shall act according to your wisdom, and don't let his gray hair go down in peace to Sheol.

7. "And show kindness to the children of Barzillai, the Gileadite, and let them be among those who eat at your table, because they were close to me when I was fleeing from Absalom, your brother.

8. "And Shimei, son of Gera, the Benjaminite from Bahurim, is here with you. And he cursed me, a sickening curse, in the day I went to Mahanaim, but he came down to me at the Jordan, and I swore to him by YHWH, saying, 'I won't kill you by the sword.' 9. And now, don't hold him innocent, because you're a wise man, and you know what you'll do to him: and you'll bring his gray hair down in blood to Sheol."

10. And David lay with his fathers, and he was buried in the city of David. 11. And the days that David ruled Israel were forty years: he ruled in Hebron seven years and ruled in Jerusalem thirty-three years. 12. And Solomon sat on the throne of David, his father, and his kingdom was very secure.

13. And Adonijah, son of Haggith, came to Bathsheba, Solomon's mother, and she said, "Do you come in peace?"

And he said, "Peace." 14. And he said, "I have a word with you."

And she said, "Speak."

15. And he said, "You know that I had the kingdom, and all Israel set their faces on me to reign, and the kingdom turned and became my brother's because it was his from YHWH. 16. And now, I ask one thing of you; don't turn me down."

And she said to him, "Speak."

17. And he said, "Say to Solomon, the king—because he won't turn you down—that he should give me Abishag, the Shunammite, as a wife."

18. And Bathsheba said, "Good. I'll speak to the king about you."

19. And Bathsheba came to King Solomon to speak to him about Adonijah. And the king got up toward her and bowed to her. And he sat on his throne and had a throne placed for the king's mother, and she sat to his right. 20. And she said, "I ask one small thing of you; don't turn me down."

And the king said to her, "Ask, my mother, because I won't turn you down."

21. And she said, "Let Abishag, the Shunammite, be given to Adonijah, your brother, as a wife."

22. And King Solomon answered, and he said to his mother, "And why do you ask for Abishag, the Shunammite, for Adonijah? Ask for the *kingdom* for him—because he's my brother who's older than I—and for him and for Abiathar, the priest, and for Joab, son of Zeruiah!" 23. And King Solomon swore by YHWH, saying, "May God do this to me and double this if Adonijah hasn't spoken this thing against his own life. 24. And

now, as YHWH lives, who established me and seated me on the throne of David, my father, and who has made me a house as he spoke: Adonijah will be put to death today." 25. And King Solomon sent by the hand of Benaiah, son of Jehoiada, and he struck him, and he died.

26. And the king said to Abiathar, the priest, "Go to Anathoth, to your fields, because you should be a dead man, but I won't put you to death today because you carried the ark of the Lord YHWH before David, my father, and because you suffered in everything in which my father suffered." 27. And Solomon expelled Abiathar from being a priest to YHWH—fulfilling the word of YHWH that He had spoken about Eli's house at Shiloh.

28. And the news came to Joab, because Joab had turned after Adonijah (but had not turned after Absalom), and Joab fled to the tent of YHWH and took hold of the horns of the altar. 29. And it was told to King Solomon that Joab had fled to the tent of YHWH and, here, was by the altar, and Solomon sent Benaiah, son of Jehoiada, saying, "Go. Strike him!"

30. And Benaiah came to the tent of YHWH and said to him, "The king says this: Come out!"

And he said, "No, but I'll die here."

And Benaiah brought word back to the king, saying, "This is what Joab spoke, and this is what he answered me."

31. And the king said to him, "Do as he said, and strike him and bury him! And you'll take the blood that Joab spilled for nothing away from me and from my father's house. 32. And YHWH will bring back his blood on his head because he struck two men who were more virtuous and better than he, and he killed them by the sword, and my father, David, didn't know: Abner, son of Ner, commander of the army of Israel, and Amasa, son of Jether, commander

of the army of Judah. 33. And their blood will come back on Joab's head and on his seed's head forever, but there will be peace forever from YHWH for David and his seed and his house and his throne."

34. And Benaiah, son of Jehoiada, went up and struck him and killed him. And he was buried at his house in the wilderness. 35. And the king put Benaiah, son of Jehoiada, in his place over the army. And the king put Zadok, the priest, in place of Abiathar.

36. And the king sent and called to Shimei and said to him, "Build a house in Jerusalem and live there and don't go out from there anywhere. 37. And it will be: in the day you go out and cross the wadi Kidron, *know* that you'll *die*. Your blood will be on your head."

38. And Shimei said to the king, "The thing is good. As my lord the king spoke, so your servant will do." And Shimei stayed in Jerusalem many days.

39. And it was at the end of three years, and two slaves of Shimei's fled to Achish, son of Maacah, king of Gath. And they told Shimei, saying, "Here, your servants are in Gath." 40. And Shimei got up and saddled his ass and went to Gath, to Achish, to seek his slaves. And Shimei went and brought back his slaves from Gath. 41. And it was told to Solomon that Shimei had gone from Jerusalem to Gath and had come back. 42. And the king sent and called Shimei and said to him, "Didn't I have you swear by YHWH and warn you, saying, 'In the day you go out and go anywhere, *know* that you'll *die*'? And you said to me, 'The thing is good. I've listened.' 43. And why haven't you kept the oath by YHWH and the commandment that I gave you?" 44. And the king said to Shimei, "You know all the bad that your heart knew, that you did to David, my father. And YHWH will bring back

your bad on your head. 45. But King Solomon is blessed, and David's throne will be secure before YHWH forever." 46. And the king commanded Benaiah, son of Jehoiada, and he went out and struck him, and he died.

And the kingdom was secure in Solomon's hand.

AFTERWORD

THE STORY BEGAN WITH THE ESTABLISHMENT OF THE EARTH AND ENDED WITH THE ESTABLISHMENT OF A KINGDOM. THAT ENDING, WHICH COMPRISES THE work's last two chapters, is so filled with the primary themes and language of the story that it is a closing showpiece of the author's artistry, weaving the many themes into a denouement. And, at the same time, it provides confirming evidence of the work's unity. Thus:

At the beginning of the story the author tells us that four rivers flow from Eden, one of which is named Gihon. We know that the choice of the Gihon was purposeful because the author makes a pun on its name: the deity's curse on the snake is: "you'll go on your belly." The Hebrew word that is translated as "belly" there is *gᵉḥōnᵉkā,* a play on Gihon. Now the story concludes with the establishment of Solomon as King David's successor, and the author tells us where Solomon is made king: at the *Gihon* (1 Kgs 1:33).

The work begins in Eden with the human becoming a "living being," endowed with the "breath of life," and it proceeds to tell the story of the tree of life and of the entry of death into the humans' world. The work's end is replete with references to life and death, from the cries of "Long live Adonijah" and "Long live Solomon" to the account of David's death, which he describes as "the way of all the earth," and Solomon's executions. The two words for "life" that are used in the beginning of the work (Hebrew *ḥay* and *nepeš*) occur ten times in these last two chapters. The word for "death" occurs eleven times.

The beginning of the work defines the relations between the sexes. It pronounces the attraction of men to women, declaring that a man will "cling" to a woman. The conclusion tells how Adonijah falls because of his attraction to a woman, the beautiful virgin Abishag, with whom David had been unable to have sex. In the ancient Near East a king's sexual ability was linked to his ability to rule. It is no coincidence that the report that David did not "know" Abishag is directly followed by the report that Adonijah declared himself king in his father's place. And when Adonijah later asks for Abishag to be given to him as a wife, whether his attraction to her is romantic or political, he is making a potential claim on his brother Solomon's throne (just as Absalom showed his claim to the throne by taking ten of David's concubines). Solomon answers, "Why do you ask for Abishag? Ask for the *kingdom!*" And Adonijah is executed.

More intriguing is the pronouncement in Eden that man is to dominate woman. There follows a long sequence of stories in which women find channels of power and influence within this male-dominated structure, culminating in the work's conclusion. There Bathsheba, as preeminent wife, acquires the succession to David's throne for her son Solomon (as Rebekah had acquired the succession for her son Jacob). In the last chapter of the work, Solomon, the ruler of an empire, rises from his throne and bows to his mother, Bathsheba. If we wonder how a female author could have written the line "Your desire will be for your man, and he'll dominate you" (Gen 3:16), the answer may lie in the author's depiction of the fate of that pronouncement, climaxing here. It reflected the realities of the author's world, in which men ruled but in which women, especially upper-class women, might sometimes achieve power through wisdom or manipulation, or through their sons.

Thus key elements that are introduced in the first chapters arrive at resolution in the last two chapters. The first chapters describe the invention of clothing; the last chapters note that David is covered with clothing to get warmth. The first chapters report that a revolving sword is placed to block the way to Eden; the last chapter reports deaths by the sword (1 Kgs 2:32; cf. 2:8). The first chapters speak of Abel's blood crying from the ground; the last chapter has five references to the blood that Joab spilled.

The account of Solomon's treatment of the rebellious Shimei at the end of the work seems to be an anticlimax, a strangely minor point on which to conclude a story of major events. But this piece, as well, connects to the opening scene of the work. The work begins with the pairing of the tree of life and the tree of knowledge of good and bad. To have one is to lose the other. The work ends with a short account of Shimei's fate, but the words "knowledge," "good," "bad," and "death" fill it, occurring fifteen times. And the formulation of the deity's command to the first human reappears here in the formulation of King Solomon's command to Shimei. God says:

> "In the day you eat from it: you'll *die!*"
>
> (Gen 2:17; repeated in 3:3)

Solomon says:

> "In the day you go out . . . you'll *die.*"
>
> (1 Kgs 2:37; repeated in 2:42)

And this formulation ("In the day you do *X* . . . you'll die.") occurs nowhere else in the Bible.

Perhaps the most dramatic of all these reminiscences and culminations is the matter of fratricide. At least seven times we have seen the issue of brother-killing-brother come up in the work. In the first chapters, Cain kills Abel. In the middle, Abimelek has seventy of his half brothers killed so that he can become a king (Judg 9:5). And the war between Benjamin and the other tribes of Israel is repeatedly, explicitly presented as an issue of brothers battling against brothers (Judg 20:13, 23, 28; 21:6). So is the struggle between Israel and Judah (2 Sam 2:26, 27). And now in the last chapter, Solomon secures his throne by having his half brother Adonijah killed. And there is a trail of connected language that shows that the fratricide stories are all intimately linked.[1]

The positions of leadership—generals, prophets, priests, kings—all arrive at some form of resolution in the last chapter. The story of kings is told, from the time when there was "no king in Israel in those days" to Abimelek's failed kingship to Saul's successful but unlasting rule to David's initiation of a dynasty, which is now fulfilled in Solomon in the last chapter. The generals who command their kings' armies are also an ongoing component of the narrative. Starting with Joab's displacing Abner over a personal vendetta, this too culminates in the work's last chapter with Benaiah's displacing and then personally executing Joab. Priesthood is traced, from the establishment of Moses' grandson, Jonathan, at Dan (Judg 18:30), to David's appointment of two chief priests, Abiathar and Zadok, to Solomon's expulsion of Abiathar in this chapter. Abiathar is identified as coming from the line of Eli, the old priest of Shiloh, and the text specially notes that his demise fulfills the prophecy that Samuel had delivered about the fall of the house of Eli. Scholars have suggested that this notice was inserted by an editor, but the evidence for the unity of this work renders that suggestion unnecessary. Moreover, the text states that Solomon "expelled" (Hebrew *wayᵉgārēš*) Abiathar (1 Kgs 2:27), a term that likewise recalls the wording of the work's first chapters, in which the deity "expelled" *wayᵉgārēš*) the humans from Eden (Gen 3:24).

The role of prophets is traced as well. The development begins with the depiction of Moses as the quintessential prophet who is at the same time the leader of Israel. The narrative then pictures Joshua as a leader who, like Moses, has direct communication from God. The stories that follow, however, picture leaders who do not themselves have such revelations: Abimelek, Jephthah, Samson, Jonathan son of Gershom. Then, with the rise of the monarchy, the prophet's role is transformed. Samuel is the last figure in the mold of Moses, a prophet who is also a leader. From the moment of Samuel's anointing of Saul, the prophet's task is to be a counterbalance to the king. It is a conflict between Samuel and King Saul that leads Samuel to a new choice: David. King David in turn must face the criticism of the prophet Nathan. It is Nathan and not David who conveys the word of God, and David is bound by it. It is Nathan who speaks in the matter of David's affair with Bathsheba. Now, in the last chapters, it is Nathan who designs the plan that leads David to choose Solomon as his successor.

These four roles—king, priest, prophet, general—are continuous elements that the author winds together like the cords of a rope all through the narrative and then brings to a combined climax: Two brothers compete to be king. Each is backed by a priest and a general: Adonijah has the priest Abiathar; Solomon has the priest Zadok. Adonijah has the general Joab; Solomon has the general Benaiah. Solomon wins because he also has the prophet, Nathan, in his party. The not-insignificant factor of his mother Bathsheba's support joins the theme of the relations between man and woman to this element of the leadership roles. And all of these elements merge in the story's finale.

There is more that culminates here. The ark containing the tablets of the Ten Commandments leads the people in the wilderness in Moses' generation. It is carried into battle during Joshua's conquests. It is captured by the Philistines and then returned in Samuel's days. David brings it to Jerusalem and houses it in a tent. And in the last chapter Joab runs to the tent and is executed there.

There is the theme of the security of the country. Promised to Abraham near the beginning of the work, it is fulfilled through Joshua near the middle. It is promised to David for his son—2 Sam 7:12 ("I'll make his kingdom secure")—and then it is achieved through Solomon at the end.

The theme of fathers and sons runs through the narrative: Noah and Ham, Abraham and Isaac, Isaac and Jacob and Esau, Jacob and Joseph, Eli and his corrupt sons, Samuel and his corrupt sons, David's struggles with his sons. In the final chapter, with the kingdom secure in Solomon's control, the first successful dynasty is established in

Israel: kingship is to flow smoothly from father to son. The symbolic moment that conveys father-son succession so powerfully in Genesis is when Jacob bows to his son Joseph on his deathbed (Gen 47:31). Now, in the concluding chapters, this act is recapitulated as David bows to his son Solomon on his deathbed (1 Kgs 1:47).

To my mind, this work is especially about families. Even when it focuses on tribes and countries, it uses familial terminology ("brothers") and makes familial connections (Num 20:14; Judg 18:14; 20:13, 23, 28; 21:6). The Ammonites and Moabites are Israel's cousins; they are the children of Abraham's nephew, Lot. The Edomites are Israel's brothers, the children of Jacob's twin brother, Esau. The first section includes the very long story of Jacob's family; the final section includes the comparably long story of David's family. And the very selection of Abraham's family in the beginning is explicitly so that they will become "a blessing to *all the families* of the earth" (Gen 12:3).

Above all, this is the story of the relations between God and humankind. It is about the balance—or the struggle—between divine direction of human affairs, on one hand, and human desire for independence on the other. At the outset, the deity is involved directly: walking in the garden, personally reprimanding the first man and woman, making their clothing, closing the ark, appearing through fire and angels. By the end, the deity is still involved but more subtly, behind the scenes. In my earlier book, *The Hidden Face of God,* I traced the progression of the gradually growing hiddenness of God and the increasing responsibility that humans must take for their destiny. There I followed this phenomenon through the course of the entire Hebrew Bible and its implications for rabbinic Judaism and the Christian New Testament. But, here, we can observe the beginning of that chain. The God who speaks to Adam, Eve, the snake, Cain, Noah, Abraham, Sarah, Isaac, Rebekah, and Jacob in Genesis is never quoted directly as speaking to anyone in the Court History.[2] And this is not a sudden difference that we might explain as reflecting a change of authorship. It is rather a gradual, continuous development through the course of the entire work, culminating, like virtually everything else, in the closing two chapters.[3] The deity has made a covenant with David promising that a human family will have continuing rule. That divine covenant is now fulfilled as the kingdom is secure in Solomon's hand. And now human kings, whom the prophet Samuel had opposed as a displacement of God, rule on earth with divine sanction.

From Gihon to Gihon, from Eden to Sinai to Jerusalem, all of these things have their denouement in these last two chapters of the Court History. But all of them started *prior* to the Court History! Most began in the J stories of Genesis. Others began with Moses, Abimelek, or Saul. *None* of these things that climax at the end of the Court History *began* in the Court History. And one cannot simply claim that an editor just rewrote these chapters and tied all of these things together here. They are too intricately wound together all through the work. Connections of this sort are what we normally point out in introductory literature courses to show students how an author fashions a text. I am pointing them out here, however, not just to display the author's artistry but to show, once more, that this is a single author's united work.

What is this work? Was it the first novel or the first history? It has qualities we associate with both. This is not a question of whether everything in it is true or not. And if it is a question of whether the *author* believed it was true—how can we know the answer to that? Though a historian can write with grace and artful wording, this first work of history is artistically formulated to a degree that we rarely—if ever—see in works called "history" as we usually mean that word. In the end, it just is what it is: a work of prose from which we can learn much about history, history writing, and literary art. As far as we know, it is the first great work of prose, and its author is the first great prose writer.

If one reads it as a collection of episodes he or she will profit but will never quite get it. It is a continuous story. If one thinks there are no continuous characters—i.e., persons who figure in the story from beginning to end—he or she is wrong. The "characters" are introduced at the beginning: God and humankind. It tells a story of relationships within human families, against the background of the relationship between God and the entire human family: "all the families of the earth." The author tells the story through the flow of generations because the author has both the instincts of the historian and the instincts of the artist. That is why arguments over whether this is literature or history are so dead-ended and unenlightening. Look at the approaches of three books that presented just the small "J" portion of the work: The first was called *The Bible's First Theologian,* the second was called *The Bible's First History,* and the third, which was called *The Book of J,*[4] was presented by a scholar of comparative literature as a literary study. All three books did much the same thing: they gave a translation of the text and made comments that were literary, historical, and theological. But the authors perceived themselves as each doing something different. I suggest that this was possible because the work in question is so very rich. The theologian, the historian, and the literary critic could all find plenty to fascinate and inspire them. And remember that the work that they were discussing was only J, which is to say, only about a fourth of the corpus!

All the approaches are valid: literary, historical, and theological. What is strange to me is when people who have chosen one of these deprecate the others: when scholars who study the Bible as literature speak of historical scholarship as passé, when scholars who look at "the text as it is" declare that scholarship that deals with authors is "nearly moribund." And, at the same time, some scholars who claim to focus on the Bible's history see the biblical authors as engaging in a kind of invention that is not legitimate history. And the theologian is stuck, trying to address theoretical, systematic questions to a text that, being a story, just is not systematic. I suspect that if we were to argue these positions in front of the author of this text, this person would be impatient or simply bored. This person wrote us a story, a long story of families that flows through generations and centuries. Historians can find a predecessor in it. So can novelists. And theologians can find inspiration.

Identifying this as a united work goes against a great deal of current biblical scholarship because so much of that scholarship is based on the acceptance of the idea that all ancient prose was short. There has been an assumption that biblical authors could not write anything long. Long works, in this view, had to be composed by unknown editors who put short works together. If I am right that this was a single long work, then this goes against all those who have identified the Joseph story as a "novella," and against all those who have seen the books of Samuel as collections of little pieces, and against scholarship on each of the sections of this work that saw collections of separate "traditions." This also goes against a great deal of current scholarship that dates these texts extremely late: at the end of the biblical period rather than the beginning. I hope that the evidence that I have presented here and in the Appendix will persuade those who have seen small pieces until now. I hope that the reading of the text of the work itself will persuade them even more. I don't know if anything at all will persuade the late daters of the Bible's story, since even the clear evidence that it was written in early Biblical Hebrew seems to have had no effect on them, and even archaeological discoveries that refer to conditions that they deny ever existed cannot change their minds. I therefore realize that this will be opposed from a variety of quarters; it will be controversial to say the least. My colleagues who are persuaded by it encourage me with their sense that this is valuable. Speaking for myself, I can only say that I came upon this work while investigating something else, and I applied a standard of doubt and a search for alternative explanations at each stage of the research. And the weight of the evidence pulled me to recognize the oneness of this work.

What are the implications of this work for our reading of the Bible as a whole? As I said at the end of *Who Wrote the Bible?,* the Bible is more than the sum of its parts. We do not have to choose between two alternatives. We can esteem this work's beauty, and we can esteem the larger work that was built out of it and around it. In its early years, the study of the Bible's authors was often an attack on the Bible, taking it apart, diminishing it. But we have reached a point at which it is the opposite: we can see the path of how the Bible came to exist. Starting with the work of a great writer, it became greater than all of its writers combined. We may value this work more for what it is itself, or we may value it more for being the foundation of the Bible. We may regard it as human or as divinely revealed or inspired. But, in any case, let us treasure it.

I have referred to this work as the heart of the Bible. To appreciate it both for what it is itself and for being the core around which the Bible was built, one must first read it and then imagine all the treasures that were added to it. Read the remarkable story of Abraham, and then add to it the account of his near sacrifice of his son Isaac, which was written by another gifted writer and later joined to this work. Add to the story of Mount Sinai the account of the golden calf. When you get over your surprise at the rituals of the Ten Commandments here, which are mostly different from the Ten Commandments that most people know, add the ethics of the version that has dominated Jewish and Christian morality. Then add the poetry of the Psalms, the wisdom of Job, and the visions of the prophets. Try to visualize the Bible growing from this root into a beautiful, multifaceted flower. Then you will understand how the Bible became the Bible—so powerful, so influential, and sacred to so many.

TEXTUAL NOTES

Gen 2:5–6 is placed between dashes and indented to convey that in the Hebrew it is a statement in the past perfect of the conditions that existed prior to the fashioning of the first human.

Gen 2:6. The translation "river" follows from P. Kyle McCarter's treatment, "The River Ordeal in Israelite Literature," *Harvard Theological Review* 66 (1973): 403–12.

Gen 2:18, 20. The identification of the woman as "a strength corresponding to him" rather than the usual "an help meet" is based on the understanding that the Hebrew term used here, *'ēzer,* can mean either "help" or "strength." As evidence of the latter, the word is found in parallel with the word *'ōz,* meaning "strength." Indeed, the name of the Judean king Uzziah (meaning "YHWH is my strength") in Chronicles is given as Azariah in Kings.

Gen 2:20. "Every beast" refers to domestic animals, as opposed to "animal of the field," which refers to wild animals.

Gen 3:20. "Eve," Hebrew *ḥawwâ,* the feminine form of the word "life."

Gen 4:2; 29:6; 30:31, 38, and numerous other occurrences. The Hebrew *ṣō'n* usually means "flock," referring to a mixture of sheep and goats. At some times, though, it must refer to sheep alone. All English translations have problems as the English terms and Hebrew terms do not match up exactly, and so it does not appear possible to be 100 percent consistent. The translator must make a case-by-case decision on which term to use.

Gen 4:8. Cain's words to Abel, "Let's go out to the field," appear in the Greek (Septuagint, LXX), but not in the Hebrew (Masoretic Text, MT).

Gen 5:29. The name "Noah" is connected figuratively here to the Hebrew root *nḥm,* meaning "console," though one would normally connect it to a different root, *nwḥ,* meaning "rest."

Gen 6:2, 4. The identity of the *b^e^nê 'ĕlōhîm* is an old and difficult problem of understanding and of translation. It may be "the sons of God" or "the sons of the gods."

Gen 7:1. The Hebrew *tēbâ* means a "box." It has long been translated as "ark," which is the Old English word for "box." Because "ark" is so old an English word, which people no longer use in daily speech and writing and because the story is so famous, people frequently think that the word really means "boat." It is not a boat in the flood story. (In the P version, its measurements are rectangular, and it has no rudder, sail, oars, or steering. Here in the J version, its measurements are not given at all.) The same applies to the box (Hebrew *tēbâ*) in which the baby Moses is placed in Exod 2:3. Still, I must admit that it is notable that the two objects called *tēbâ* in the Hebrew Bible both float on water. I have used the word "ark" here reluctantly because it is now so familiar that readers would find any other word jarring, and it might distract them from the flow of the story.

A different Hebrew word, *'ārôn,* on the other hand, is used for the box that contains the tablets of the Ten Commandments. "Ark" is still an appropriate translation for this container (1) because people know what it is and would not confuse it with a boat; and (2) because that word is still used for the box that contains the scrolls of the Torah in synagogues to this day and is therefore not archaic English in the way it is in the case of Noah's "ark."

Gen 10:8–30. The genealogical material is difficult to separate into sources and identify. Though much of it has frequently been identified as J, I have taken the most limited path and left it out of this translation.

Gen 11:9. The name Babylon, Hebrew *bābel,* is connected figuratively here to the Hebrew word *bābel,* meaning "mixed up" or "confused." I used the English word "babbled" in order to convey the pun in the Hebrew.

Gen 12:1. No translation quite captures the sense of the Hebrew *lek l^{e}kā* ("Go you," "Get you," "Go for yourself"). The second element, *l^{e}kā* (or *lākem),* occasionally follows verbs in this work. For example, in Num 14:25 we have *s^{e}'û lākem* ("Travel you," "Travel for yourselves"); and in Josh 7:10 we have *qûm lāk* ("Get you up," "Get up for yourself"). I believe it is better to use no English term than to use any of the possible equivalents, all of which are clumsy English and thus misrepresent the original, which is elegant Hebrew.

Gen 12:1, 11. The idea that the name Abraham was originally Abram and was later changed to Abraham occurs only in P (Gen 17:5), not in J. The same applies to the name of Sarai, changed to Sarah, in P (17:15). In this work, therefore, I understand the names to have been Abraham and Sarah throughout. They now appear in the Bible as Abram and Sarai prior to the P account of the changes of their names presumably because an editor (R?) decided to make the name change chronologically consistent at the time of the combining of the sources and substituted Abram for Abraham in the early part of J. Similarly, the change of Jacob's name to Israel occurs in E and P, not in J. The name Israel thus appears for the first time in J in Gen 34:7 and 35:21–22, seemingly out of nowhere. Again, the problem appears to be the result of the combination of J with E and P.

Gen 15:7. This verse reads, "I am YHWH, who brought you out of Ur . . ." There is a well-known conflict in the texts as to whether Abraham set out from Ur or Haran, a conflict that already troubled the medieval commentators (see Rashi and Ramban). In P, Abraham's father first takes the family from Ur to Haran, and Abraham subsequently goes from Haran to Canaan (11:31f; 12:4b, 5). In J, God tells Abraham to "go from your land and from your birthplace" (12:1). This place is not named in J. When Abraham later sends his servant back to "my land and my birthplace" (24:4), the servant goes to "the city of Nahor," which turns out to be *Haran* (Gen 24:10ff; 27:43; 28:10). These facts suggest that the original reading of Gen 15:7 would have been "Haran" and that the Redactor who combined J with P changed it to "Ur" to make it consistent with what precedes.

Gen 16:13. (1) "El-roi," meaning "God who sees me."

(2) "Have I also seen after the one who sees me here?" This is a difficult clause in the original; all translations of this clause are uncertain.

Gen 18:2; 19:1; and many more occurrences: "bowed to the ground," "he bowed, nose to the ground." The Hebrew verb means to bow prostrate.

Gen 19:22. The name "Zoar" has the same root as the word "small," which Lot uses twice in his plea to go there (19:20).

Gen 24:33. The word "bread" appears in the LXX but not in the MT.

Gen 25:23. On the subtle double meaning in the wording of the oracle to Rebekah, see my *The Hidden Face of God* (San Francisco: Harper San Francisco, 1995), pp. 111–12.

Gen 25:26. The name "Jacob" is connected here to the Hebrew word for "heel."

Gen 25:30. The name "Edom" is connected here to the Hebrew word for "red," *'ādōm*. Esau is identified as the eponymous ancestor of the Edomites.

Gen 26:20. The name "Esek" is connected here to the Hebrew word for "tangled."

Gen 26:22. The name "Rehovot" is connected to the Hebrew root meaning "to extend." J appears to play on this root six times, while it never occurs in E, which is interesting because it is the root of the name Rehoboam, the first king of Judah after its separation from Israel. (See the Appendix, p. 341.)

Gen 27:5. The words "to his father" appear in the LXX. They appear to have been omitted by the scribe because of their similarity to the preceding word in the Hebrew, thus: *lhby'* ("to bring") *l'byhw* ("to his father").

Gen 27:28, 39. David Noel Freedman notes that there is play on words here, as the blessings of the two sons hinge on the term *min.* In the blessing given to Jacob he is promised access to and a rich share of the rain and the dew, with *min* being used partitively. In Esau's blessing it is used spatially to mean that Esau will be denied access to fertile lands and the fruits of the soil. So what was a blessing for Jacob becomes in effect a kind of curse for Esau, and the words are practically identical.

Gen 27:36. Esau now makes a new play on the name "Jacob" (connected earlier to the word for "heel" in Gen 25:26 above), connecting it to a verbal meaning of "to usurp."

Gen 28:19. Hebrew "Beth-El" means "House of God."

Gen 29:32–35. The four sons' names are each connected with a Hebrew verb: Reuben ("to see"), Simeon ("to hear"), Levi ("to bind"), Judah ("to praise").

Gen 30:24b. The name "Joseph" is connected here to the Hebrew root *ysp,* meaning "to add." (In E Joseph's name is traced to a different root, *'sp* in Gen 30:23–24a.)

Gen 32–33. The account of the reunion of Jacob and Esau presents some of the most difficult problems for *all* views of authorship known to me. In *Who Wrote the Bible?* ([San Francisco: Harper San Francisco, 1987], p. 257), I noted this and cautioned that some of my identifications of J and E are tentative. I have identified the entire reunion scene in 33:1–17 as E, which means that whatever culmination there was in J of the conflict between Jacob and Esau is now missing. Presumably the redactor who combined J and E (RJE) favored the E account and either eliminated the J account entirely or worked in only small portions of it. I have retained a few verses from the reunion scene that could possibly have been worked in from J, but I acknowledge that this is tentative.

Gen 34:7; 35:21f. The origin of the name Israel is not told in J. (It appears as a story of a change of Jacob's name to Israel in E in Genesis 32, and in P in Genesis 35.)

Gen 34:14. "A man who has a foreskin" is almost always translated "a man who is uncircumcised," presumably because the translators find the original too crude, and so they soften it in translation. I may understand their imposition of their taste on the text, but I felt obligated to retain the force of the original language. (I use "uncircumcised" in a few cases where the context requires it.) Similarly with the expression "anyone who pisses against the wall"—meaning all males—in 1 Sam 25:22, 34, I have translated the original literally rather than replacing the author's earthy original with some English euphemism.

Gen 36:31–39 contains a list of "the kings who ruled in the land of Edom before a king ruled the children of Israel." I listed it among the passages of this work on p. 12, but I have not included it in the translation because I am not sure where it originally went. It seems out of place here between the story of Dinah and the story of Joseph. It may originally have come after the parting of Jacob and Esau at Gen 33:16. In any case, it is significant because it makes a firm connection between the time of the patriarchs and the time of the birth of the monarchy, which is to say a connection between J and the Court History. It is also significant because it is one more element that fits with the tie between J and the kingdom of Judah. J, coming from Judah, which bordered on Edom, includes this detailed interest in Edom's kings. E, coming from Israel, which did not border on Judah, has no such text as this.

Gen 38:29. The name "Perez" is connected here to the Hebrew root *prṣ,* meaning "to make a breach."

Gen 43:28. The words "And he said, 'Blessed of God is that man'" appear in the LXX and Sam. This plus presumably was lost from the

MT as a result of haplography when a scribe's eye jumped from the *wy* at the beginning of *wayy'ōmer* to the beginning of *wayyiqq^{e}dû*.

Genesis 49. This poem, known as the Blessing of Jacob, was not composed by the author of the prose work. It may very well have been the author's source for the matter of the demotions of Reuben, Simeon, and Levi. In any case, the author worked it into the narrative in such a way that it now provides the denouement of these and other matters.

Gen 49:25–26. David Noel Freedman proposed this reconstruction. Thus:

> *brkt šmym m'l*
> *brkt thwm rbṣt tḥt*
> *brkt šdym wrḥm*
> *brkt 'byk gbr w'l*
> *brkt hwry 'd*
> *t'wt gb't 'wlm*

Exod 2:10. The name Moses is connected here to the Hebrew root *mšh,* meaning "to draw."

Exod 2:22. The name of Moses' son, Gershom, is connected here to the word *gēr,* meaning an alien.

Exod 4:20. In J, Moses has only one known son, Gershom. The addition of a single letter here to make it "sons" appears to be the work of the redactor who combined J with E (RJE) in order to reconcile this passage with the E report that Moses had two sons, Gershom and Eliezer, in Exod 18:3–4, which contains a doublet of the naming of Gershom.

Exod 4:24–26. The episode at the lodging place is one of the most difficult passages in the Torah to understand and translate. No solution known to me seems adequate. I translated v. 24 as "he [meaning Moses] asked to kill him" rather than the usual "He [meaning God] sought to kill him" in order to raise an alternative possibility, namely, that Moses, who is fearful and reluctant, asks/seeks an end to his life rather than to go on to Egypt. I admit that this possibility has its share of uncertainty as well. The idea that Moses would ask for his own death is quite conceivable, in any case, because he makes such a request in a story in E (Num 11:15).

Exod 5:3–12:51. The plagues narrative appears to me to be composed of P and E. It is possible that some lines from J have been mixed into the E narrative, but I am not able to identify them. My colleague William Propp and I each came to this conclusion, that the non-P version of the plagues account is E, not J, independently. See his explanation of this in his commentary on Exodus in the *Anchor Bible*.

Exod 15:1–18. The Song of the Sea is an independent composition, older than this work, but it may have been incorporated in the work.

Exod 19:24. The reference to Aaron's coming up with Moses in this verse is strange because: (1) Aaron has not been introduced in this work up to this point. (2) Aaron does not in fact come up with Moses; Moses is told explicitly that no one is to come up with him (34:3). (3) Aaron is not mentioned in the story in this work at all outside of this verse. Aaron's name does not come up after this until Judg 20:28, and then again in 1 Sam 12:6. I therefore take the reference to him here in Exod 19:24 to be a probable addition by the redactor (R).

Exod 34:1b, 4. The references to an earlier set of tablets that were broken appear to have been added by the redactor who combined J and E (RJE) in order to include both the E golden calf story and the J story of the Ten Commandments.

Exod 34:14–26. The first, second, and fourth of the Ten Commandments are similar to those in the more familiar versions of the Decalogue, which appear in Exodus 20 and Deuteronomy 5. The rest of the J version will come as a surprise to many readers as it concentrates on ceremonies and lacks the prohibitions of murder, theft, and adultery. Scholars have frequently distinguished between the J list as the "ritual" Ten Commandments and the other two (Exodus 20 and Deuteronomy 5) as the "ethical" Ten Commandments.

Num 13:17. The text says, "And Moses sent *them.*" The original object of "sent" here was apparently changed when the Redactor combined this story with the P story of the spies. The P report of a divine command to send the men was placed before the start of the J story. This report of the sending in v. 17 is the Redactor's repetition of the report that already exists in the P text in v. 3. Such a repetition that occurs precisely at the point where two sources are joined is known as a "resumptive repetition," or an "epanalepsis."

Numbers 22–24. I have not included the Balaam narrative in the work because I have never felt I could confidently separate J and E components of the story. This has proven to be one of the most difficult of passages, and practically everyone has found it to be extremely complex. In *Who Wrote the Bible?* I simply placed it all in the E column, but that was just a recognition that I was not prepared to take on the long passage with all its complexity at that time.

Joshua 1. It is difficult to separate the first chapter of Joshua into sources. It appears to be a composite introductory section composed by the Deuteronomistic historian, using one or two sources and then adding some wording of his own, notably vv. 7–9a. I have selected a minimum of lines from the chapter, recognizing that the transition between Moses' death and Joshua's sending the spies from Shittim is probably embedded in this chapter. Other selections are possible. The main point is that Joshua 2 begins in the place in which J in the Torah leaves off and that one need not imagine any wording beyond what is contained in Joshua 1 as being the bridge between the two passages.

Josh 3:16. "The Salt Sea" is still known in Hebrew by that name: *yam hammelaḥ*. It is the body of water now known in English as the "Dead Sea."

Joshua 4. It is difficult to determine what portion of this chapter belongs to the work and what has been added to it, and there is no characteristic language to help make this judgment. I have therefore not included it in the translation.

Josh 6:17 and elsewhere. "Complete destruction." Hebrew *ḥerem* refers to the rule in divinely commanded wars of not taking spoils or slaves—the war is not for profit—but rather destroying all of these and thus dedicating them to the deity. In Josh 7:1, I translate it as "the rule of complete destruction."

Josh 7:20. "Sinned." It has become almost a commonplace to say that the Hebrew root *ḥt'* does not mean "to sin," but rather "to err," in the sense of "to miss the mark." But the word appears to have this meaning in only one place, Judg 20:16, where the text reports that the Benjaminite warriors "could sling a stone at a hair and not miss." Elsewhere the word is grouped with other terms that denote the category of transgression, offense, and crime, such as *'āwōn* and *peša'* (Exod 34:7).

Josh 8:1. The words "Don't be afraid and don't be dismayed" are usually regarded as Deuteronomistic; but, as in the case of Joshua 1 and of 2 Samuel 7 (see the notes there), we cannot be sure whether the Deuteronomistic historian learned them precisely from the source here. This may apply as well to other expressions that are usually credited to the Deuteronomistic historian.

Josh 9:16. The words "at the end of three days" are not part of this work. They are from another source, P, which has been combined with J in this chapter.

Josh 12; 13:1–13; 14:6–15; and portions of **23** and **24** are difficult to identify as presently mingled with other sources. This makes it difficult to tell how the account of Joshua ended in this work and whether it was connected to the beginning of the account in Judges. The epanalepsis of the words "And the land had respite from war," which occur in 11:23 and then recur in 14:15, suggests that some portion of the intervening material was secondarily added to the text. I have included only 13:1–13 here, mainly because it includes a report of Joshua's old age that fits with the language and context of the work.

Judg 8:30–32. The story of Abimelek, son of *Jerubbaal,* begins in Judg 9:1. A previous story explains that Jerubbaal and Gideon are two names of the same man (6:25–32; 7:1), but that previous story is not part of this work. I therefore take the references to Abimelek's father as Gideon in 8:30–32 to have been made by the editor (known as the Deuteronomistic historian) who connected that source to this work, and I read Jerubbaal here.

Judg 11:35. The verse is an embroidery of puns: "When he saw her" (כראותו); "and he ripped" (ויקרע); "you've brought me down" (הכרע הכרעתני); "among the ones who anguish me" (בעכרי).

Judg 11:40. "Regularly" here and elsewhere translates the Hebrew phrase that literally is "from days to days," which is usually understood to mean "annually." That may well be correct, but since I am not certain I have used the less specific "regularly."

Judg 12:1–6. In terms of continuity, this passage may or may not belong to this work. It is followed by a list of judges that has no connection with this work, and it does not contain any of the characteristic language. I have not included it in the translation.

Judg 13:4, 7. "Beer." I am told that they did not have distilled alcohol in the ancient world and therefore no hard liquor. The alcoholic drinks were wine and beer.

Judg 14:3. The Greek has "your people." The MT has "my people." "Your" makes better sense in the context.

Judg 16:13–14. A long portion of these two verses appears in the Greek but not in the MT, namely: "'and fasten it with the pin into the wall, then I'd weaken and be like any man.' And it was when he was asleep: and Delilah took seven locks of his head and wove them into the web." It appears to be a classic example of haplography, in which this portion was left out because the MT scribe's eye jumped from the word "web" to the next occurrence of the same word.

Judg 17:5. An ephod is a sacred garment generally worn by a priest (1 Sam 2:28; though David wears an ephod in 2 Sam 6:14). Teraphim are small idols.

Judg 18:22. "Micah and the people." Reading with the Greek.

Judg 18:30. (1) "Jonathan son of Gershom son of *Moses.*" This is a classic and important problem of text criticism. The MT has *m*^{n}*śh,* with the letter *n* (Hebrew nun) suspended above the line. The question is whether the original text read Manasseh *(mnśh)* or Moses *(mšh).* The Greek text has Moses (Mousas). In J, there is a record of only one son born to Moses: Gershom (Exod 2:22). It is in E that Moses has another son, Eliezer (Exod 18:3–4). When we now eliminate all the intervening material and read this work on its own, (a) Moses' other son is eliminated from the account, and (b) the Judges text moves closer to the accounts of Moses, and so this connection of the priest in Judges to the line of Moses becomes even stronger than before: he is Moses' grandson. This in turn strengthens the case concerning the existence of a priesthood whose members traced their descent from Moses in ancient Israel.

(2) The verse concludes that the Danite priesthood functioned "until the day of the land's exile." Since other evidence indicates that this work was composed before the exile of the land of Israel by the Assyrians in 722 B.C.E., this reference to "the day of the land's exile" must be traced to the Deuteronomistic historian. The most likely sce-

nario is that the original text said "unto this day," and the historian (of Dtr^{1}) changed the wording to fit the historical situation that prevailed after the destruction of the northern kingdom of Israel.

Judg 20:16. The reference to the binding of the right hand of the Benjaminite soldiers suggests a military practice of deliberately tying up the stronger hand of most people so as to force them to use their left hand more and thus become ambidextrous. This training technique was known among the Maori as well. This expression has generally been translated "left-handed," which makes little sense. Baruch Halpern makes the connection to training for dexterity in *The First Historians* (San Francisco: Harper San Francisco, 1988), pp. 40–43.

Judg 20:27–28. "And the ark of God's covenant was there in those days. And Phinehas, son of Eleazar, son of Aaron, was standing in front of it in those days." These words may not originally have been part of the work. The work has few, if any, certainly original references to Aaron; and this line certainly appears to be intrusive in the context.

1 Sam 1:20. The name "Samuel" is connected here to the Hebrew root *š'l,* meaning "to ask," which occurs seven times in this chapter. Since it is actually the root of the name Saul and not Samuel, many scholars have assumed that this story was originally about Saul and was misplaced here. But that assumption implies that the author or editor of this story was inexplicably clumsy in Hebrew. On the contrary, we should note that several of this author's etymologies are based on similarity of *sounds,* not on root meanings. Thus this author connected the name Noah to the root *nḥm,* "to console" (see above, Gen 5:29). The extra *mem* in the name Samuel need not have bothered the author any more here than it did there in the case of Noah. (See Appendix, p. 341.)

1 Sam 1:24. The verse should be read *"to Shiloh* when she had weaned him, with a *three-year-old bull"* (rather than the "three bulls" in the MT) based on the Qumran and Greek texts and on the reference to a single bull in the next verse.

1 Samuel 2–3 recounts the subordination of a priesthood that formerly was preeminent to another—which turns out to be the subordination of Abiathar to Zadok, which is then made explicit in 1 Kings 2:27. For those of us who see the Deuteronomistic historian as connected to the non-Zadokite (Mushite) priesthood, the question has been: why would someone from that community develop this story that is critical of his own priestly ancestors? The solution lies in the recognition that this story does not come from the historian himself but from his source. Given the continuity of this source as it appears in the Deuteronomistic history (as opposed to its fragmented condition in Genesis through Numbers), it appears that the Deuteronomistic historian retained it virtually in toto, adding material to it but not feeling free to delete from it, even material that criticized one of his own group's ancestors.

1 Sam 2:25. I have tried to convey the Hebrew pun here *(pllw* and *ytpll lw)* with English "appraise" and "prays."

1 Sam 2:33. The word for "by the sword of" occurs in a Qumran text.

1 Sam 11:1. "And Nahash the Ammonite . . ." It appears that a scribe omitted an entire paragraph here by haplography because the paragraph began with the words "And Nahash," and the scribe's eye skipped down to the next occurrence of these words. We have the paragraph in a Qumran scroll (4QSam[a]). Frank Moore Cross translates it as follows:

> And Nahash, king of the Ammonites, sorely oppressed the children of Gad and the children of Reuben, and he gouged out all their right eyes and struck terror and dread in Israel. There was not left one among the children of Israel beyond Jordan whose right eye was not gouged out by Nahash the king of the children

of Ammon, except seven thousand men [who] fled from the children of Ammon and entered Jabesh-Gilead. About a month later Nahash the Ammonite went up . . .

See Cross, "The Ammonite Oppression of the Tribes of Gad and Reuben: Missing Verses from 1 Samuel 11 found in 4QSamuel[a]," in H. Tadmor and M. Weinfeld, eds., *History, Historiography and Interpretation* (Jerusalem: Magnes, 1997), pp. 148–58; and P. Kyle McCarter's *I Samuel* in the *Anchor Bible*.

1 Sam 15:27. The name Saul is included in the Qumran text and the Greek.

1 Sam 17:4. "A man of the land between" is apparently a technical term for the no-man's-land between the two camps. When a champion steps into that disputed zone, it throws down a challenge to the opposing side.

1 Sam 25:22, 34. The expression "anyone who pisses against the wall" is an earthy way of saying "all males." See the note on Gen 34:14 above.

1 Sam 25:30. Hebrew *nāgîd* is translated here as "designated king." Baruch Halpern (*The Constitution of the Monarchy,* Harvard Semitic Monographs [Atlanta: Scholars Press, 1981], chap. 1) has shown that Hebrew *nāgîd* refers to the person who is divinely designated to be king, as conveyed by a prophet. (On one occasion a human designates the king—David designates Solomon—but this is still understood to be divinely derived; it involves a role played by the prophet Nathan.)

2 Sam 1:18. I do not know the meaning of the enigmatic occurrence of the word "bow" in this verse. Perhaps it was the title of the song, or perhaps it was known as the "The Bow Song." I have not attempted to translate it here.

2 Sam 2:8; 9:6; 11:21. The names Ishbosheth, Mephibosheth, and Jerubbesheth were originally Ishbaal, Mephibaal (or Meribaal), and Jerubbaal, as they appear in the Greek and in Judges and Chronicles. These names come from a period during which it was possible to use the term *ba'al,* which means "Lord," for YHWH. Later, this term was differentiated from *'ădōnāy,* the other term for "Lord," so that *ba'al* was used for the Canaanite deity and *'ădōnāy* was used for YHWH. A later editor substituted the deprecating term *bōšet,* meaning "shame," for *ba'al.*

2 Sam 3:34. From Qumran we have a missing word, "manacles," supplied to parallel the "shackles" of the second half of the line.

2 Sam 6:2. "Ark which is called 'Name.'" The Hebrew and Greek differ here, and the Hebrew and the Greek of the parallel passage in 1 Chr 13:6 complicate the matter further. The key portion is lost in the Qumran fragment (4QSam[a]; see McCarter, *II Samuel, Anchor Bible* [Garden City, NY: Doubleday, 1984], p. 163).

The Hebrew of 2 Sam 6:2 reads literally:

> the ark of God which is called name name of YHWH
> of hosts who is enthroned on the cherubs on it

The Hebrew of 1 Chr 13:6 reads:

> the ark of God YHWH who is enthroned on the
> cherubs which is called name

A scholar's first impulse is to take the repetition of the word "name" in 2 Samuel as a scribe's dittography and to emend the text. But this does not account for the strangeness of the other three versions. The key to all of them may be that the ark was formally titled "The Name of YHWH of Hosts Who Is Enthroned on the Cherubs," or just "YHWH of Hosts Who Is Enthroned on the Cherubs," and that the brief form of this appellation that people used was "Name" (similar to the way contemporary orthodox Jews say *"hashem"* rather than pronounce the name of God). That is, the ark was regarded as the dwelling of the divine name. Thus, note that the Deuteronomistic formulation "the place where YHWH causes His name to dwell" refers always to the place where the ark is located.

2 Sam 6:5. Here, too, the MT and the Greek of Samuel and of Chronicles present a picture of a complex scribal history. The consonants are:

2 Sam 6:5	*bkl ʿṣy brwšym*	"in all the cypress trees"
1 Chr 13:8	*bkl ʿz wbšyrym*	"with all their might and with songs"

(Another possibility is *bkly ʿz wbšyrym,* referring to musical instruments.)

2 Samuel 7. The oracle through the prophet Nathan in this chapter contains the establishment of the Davidic covenant. As I have discussed elsewhere, the formulation of this covenant is of major importance to the Deuteronomistic history. It is difficult to determine which portions of the oracle were originally part of this work and which were added by the Deuteronomistic historian. Even if a word or phrase from this passage also occurs elsewhere in the Deuteronomistic history, that does not mean that the Deuteronomistic historian added it here. The explanation may also be that the Deuteronomistic historian found it here in one of his sources and chose to use it in other parts of his history. I have therefore retained a fairly large portion of the oracle here. I recognize that in this chapter the Deuteronomistic historian's intertwining of his own writing with the words of his source text may be quite complex. (For an analysis of this problem and citations of scholarly treatments, see McCarter, *II Samuel, Anchor Bible,* pp. 215–24.)

2 Sam 12:14. So as not to write the words "you rejected YHWH," the words "the enemies of" are understood to have been added to the phrase (either by the author or by someone else: a scribe or the Deuteronomistic historian).

2 Sam 15:7. Some texts say four years, and some say forty.

2 Sam 15:12. The Greek adds "summoned," which is not in the MT.

2 Sam 18:3. "You are like . . ." The Greek reads *'th,* "you." The MT reads *'th,* "now."

2 Sam 21. The relationship of 2 Samuel 21 to the rest of the work is not clear. It refers to matters that occur in Samuel A rather than Samuel B, specifically: Jonathan's oath to David and the marriage of Saul's daughter Merab (written in most texts as Michal) to Adriel. It also refers to the defeat of Goliath by Elhanan rather than by David, which runs contrary to both Samuel A and Samuel B. It identifies the Gibeonites without reference to their identification in Joshua. And it refers to a story of Saul and the Gibeonites that does not appear in Samuel A or Samuel B. I therefore did not include it in the text here.

1 Kings 1:1–5. It is not an accident that the story of Adonijah's attempt to take the throne comes immediately after the report that David could not get warmth with a beautiful woman. A king's sexual potency was directly related to his acceptance as ruler. Thus Absalom began his claim on the throne by having sex with David's concubines. Thus David's impotence is grounds for Adonijah to claim the throne. Thus Adonijah's request for Abishag—the woman with whom David could not perform—later will be seen by Solomon as a claim on the throne, resulting in Solomon's execution of his half brother, Adonijah.

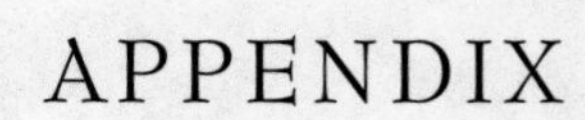

APPENDIX

THE PURPOSE OF THE APPENDIX IS TO PROVIDE A MORE DETAILED TREATMENT OF THE EVIDENCE FOR THE UNITY AND ANTIQUITY OF THE WORK THAT I have attributed to the first great writer of prose. It will be of most use to biblical scholars, but I have tried to formulate this treatment in language that is also accessible to interested laypersons. I have tried to minimize the use of technical terminology. The Appendix has four sections. The first is a collection of evidence for the work's unity. The second contains evidence for its antiquity. The third is a response to and criticism of recent scholarship that asserts a late date of composition of these texts. And the fourth is a chart showing the distribution of terminology that characterizes this work.

I

CONVERGING EVIDENCE FOR THE UNITY OF THE WORK

As I described in the Introduction, the strength of the case for the unity of this work from Genesis 2 to 1 Kings 2 is the convergence of several distinct lines of evidence. The converging lines of evidence are as follows:

I. Terminology

There is a bank of terminology that lines up uniquely in this particular group of texts. It includes both single words and phrases, and its frequency and consistency rule out coincidence as an explanation. And it crosses too many lines of genre and subject matter to be explained as deriving from mere convergence of such things. Here is a sampling of the cases:

1. I have noted the similar use of the term "foolhardy thing" (Hebrew *n*[e]*bālâ*) in the Tamar and the Dinah accounts; and the pun on this root in the Court History's story of Abigail and Nabal is well known: Abigail says of her foolish husband, Nabal, "As his name is, that's how he is!" In fact, there are ten occurrences of this term in all of biblical prose. And all ten are in this group of texts.
2. I also noted that deception is a recurring theme in J and in the Court History, but it is also articulated explicitly in the language of the texts. Jacob says to Laban, "Why did you deceive me?" (למה רמיתני). So says Joshua to the Gibeonites, Saul to Michal, and the woman of En-dor to Saul. Of seven occurrences of למה רמיתני and במרמה in biblical prose, all are in this group of texts.
3. Of seven occurrences of the expression "to wash the feet" in biblical prose, all are in this group.
4. Six of seven occurrences of the word "roof" *(gāg)* as a specific plot setting (and not just a general reference to roofs) are in this group.
5. All references to Sheol in biblical prose (nine) are in this group of texts.
6. All nine occurrences of the term for shearing *(gzz)* are in this group.
7. The phrase for "those who live in the land" *(ywšb h'rṣ)* applies to the Canaanite inhabitants of the land but refers to them in the singular. This formulation occurs six times in biblical prose. All six are in this group (Gen 50:11; Exod 34:12, 15; Num 14:14; Judg 11:21; 2 Sam 5:6).
8. J plays upon the Judean clan name Perez in Genesis 38. The verb occurs thirteen times in this group, but only once in E. Otherwise it occurs in the Hebrew Bible only in another concentration in the Chronicler's history (seventeen times) and twice in Kings.

9. The term for keeping quiet *(hḥryš)* occurs twelve times, and nine are in this group.
10. Both cases of the "coat of many colors" *(ktnt psym)* are in this group.
11. All three occurrences of the expression "weak eyes" *(khth ʿynw)* are in this group.
12. The expression for deep sleep, "a slumber descended" *(trdmh nplh)*—three occurrences, all in this group.
13. The expression "to be content to live" occurs four times in the Hebrew Bible, referring to Moses (Exod 2:21), the Israelites (Josh 7:7), the Levite of Dan (Judg 17:11), and the Levite of the mountains of Ephraim (Judg 19:6). All are in this group.[1]
14. The term for old age *(zqn bʾ bymym)* is applied to Abraham, Isaac, Joshua, and David, all in this group.
15. The term for water drying up *(ḥrb)* is consistently the same in this group: for the drying of the waters after the flood (Gen 8:13; cf. 7:22), the drying of the Red Sea during the Exodus (Exod 14:21), and the drying of the Jordan in Joshua (Josh 3:17). (Compare P, which mixes terms, using *ḥrb* and *ybš* in the flood story and *ybš* in the splitting of the Red Sea.)
16. The expression for opening the mouth *(pṣth pyh)*—four occurrences, all in this group.
17. The expression "to turn *(nṭh)* to the right or the left"—four occurrences, all in this group.[2]
18. The term for animal feed (fodder, *mśpwʾ*)—five occurrences, all in this group.
19. The expression "flesh and bone" *(bśr wʿṣm)*—seven occurrences, six in this group.
20. The term "good-for-nothing" *(blyʿl)*—eleven of fourteen occurrences are in this group.
21. The root *bśr,* "to bring news"—eleven of twelve occurrences.

22. The root *śn',* "to hate," appears seventeen times in biblical prose, thirteen in this group.
23. The term "to lie with" *(škb)* with sexual connotation—thirty-two cases in biblical prose, thirty of them in this group.
24. The expression "faithfulness and truth" *(ḥesed wā'ĕmet)*—seven occurrences in biblical prose, all in this group.
25. The word "spies" *(mrglym)* occurs twelve times in the Hebrew Bible, all in this group.

A chart showing all of the occurrences of these twenty-five terms and phrases plus twenty-five more appears on pp. 379–387.

II. Narrative Continuity

There are visible elements of narrative continuity through this corpus, linking the very texts in which the characteristic words and phrases occur:

In the last J passages in the Pentateuch, Israel is located at Shittim (Num 25:1–5); and that is where Joshua is at the point where the lexical affinities begin at the beginning of the book of Joshua, when he sends his spies to Jericho (Josh 2:1; 3:1).

The beginning of Samuel B connects to the conclusion of Judges. Geographically, Judges 21 ends with the taking of wives from Shiloh, and Samuel B begins in Shiloh (1 Samuel 1). Even more indicative of the narrative continuity: the account in Judg 11:40 reports that Israelite women would go out "regularly" (מימים ימימה) to commemorate Jephthah's daughter. It later reports in Judg 21:19 that the wifeless Benjaminites captured women who went out on the occasion of the regular holiday at Shiloh: "Here's a holiday of YHWH in Shiloh regularly" (הנה חג ה' בשלו מימים ימימה). And then Samuel B begins with the notation that Elkanah would go up to sacrifice "regularly at Shiloh" (מימים...ימימה בשלה) (1 Sam 1:3; 2:19). And these are the only occurrences of this expression in biblical narrative.[3]

Samuel B, meanwhile, flows integrally into the Court History, which depends upon it on several important points, including a number of the central players in the story. Samuel B includes the introduction into the narrative of David's wives and his generals (the Seruyah family). It introduces the matter of the relationship between Jabesh-Gilead and Saul (1 Samuel 11). And it follows the line of the priestly house of Eli at Shiloh, whose downfall is predicted in 1 Samuel, and this prediction is fulfilled in the Court History with the banishing of Abiathar.

There are other elements of narrative continuity running through this collection:

Josh 13:13 notes that "Geshur and Maachat [cf. J (Gen 22:24) referring to Maachah] have lived among Israel to this day." Geshur is in fact identified twice, in the first and last verses of the Joshua 13 list, as having continued to reside in Israel (13:2, 13). Samuel B then tells that David raids the Geshurites (1 Sam 27:8); and the Court History relates that David marries a Geshurite princess, Maachah (2 Sam 3:3), and that Absalom, the product of that union, flees to his grandfather, the king of Geshur, in the wake of the Amnon-and-Tamar episode (2 Sam 13:37f.; 14:23, 32; 15:8).

The matter of the ark is also a continuing narrative concern in this collection. The iconography of the Pentateuchal sources reflects the time and locale in which each was produced.[4] Notably, the ark, which was located in Judah, is referred to in J (Num 10:33, 35; 14:44), but never in E. The Tabernacle, which is supposed to have stood at Shiloh in the north but never in Judah, is referred to in E but never in J.[5] The importance of the ark in J continues in Joshua (chaps. 3; 4; 6; 7; 8), in a passage in one of the identified chapters of Judges (20:27), in several places in Samuel B (1 Samuel 3; 4; 5; 6; 7; but only once in Samuel A), and several times in the Court History (2 Sam 6; 7:2; 11:11; 15:24, 25, 29; 1 Kgs 2:26). And the connection is underscored by another case of a repeated prose image: when the ark is carried around Jericho, the people "shouted a big shout" (Josh 6:5, 20); and when the ark is carried into an ill-fated battle against the Philistines in Samuel B, the people "shouted a big shout" (1 Sam 4:5).

In J Abraham invokes the name YHWH for the first time at a site that has "Beth-El to the west and Ai to the east" (Gen 12:8). He also builds an altar there. The site is noted again in a report that Abraham returns there and invokes the divine name again. There is no explanation of the significance of this place between the two cities and no other occurrence of this location in J in the Torah. In the book of Joshua, though, in the section that bears the characteristic terminology that we have seen, the location of the ambush by which the deity delivers Ai into the Israelites' hands is "between Beth-El and Ai, to the west of Ai" (Josh 8:9, 12; cf. 12:9). It appears that Abraham's pious actions in Genesis have a consequence in Joshua, and this happens within the collection that we are observing.

There is also the matter of giants, which I discussed in the Introduction. Their origin is explained in the J story of the "sons of God" who have relations with human women and give birth to giants (Gen 6:1–4). Then in the J version of the spies episode, the Israelite spies see the giants (Numbers 13). Then Joshua eliminates the giants from all of the land except from the Philistine cities of Gaza, Ashdod, and *Gath* (Josh 11:21–22). And then in Samuel B the Philistine giant Goliath comes from Gath (1 Sam 17:4). (But in the Samuel A version, which is a doublet, Goliath is not described as being big [1 Sam 17:23]).

And of course there is the matter of the Abrahamic covenant, related in a J passage in Genesis 15 and fulfilled in David in 2 Samuel.

The point is that (1) there is narrative continuity and (2) this continuity falls along the lines of the characteristic terms and phrases that mark these texts.

III. Allusion

Complementary to these cases of narrative continuity is another line of evidence: cases of allusion. Passages in this collection refer back explicitly to passages that come in earlier sections:

As I noted in the Introduction, the account of the mass circumcision of the Israelites in Joshua alludes to the spies episode in Numbers, and it refers explicitly to the words of the J version of that story. In J (Num 14:22f.), YHWH swears concerning those who did not listen to his voice that they will not see the land:

...ולא שמעו בקולי אם יראו את הארץ אשר נשבעתי לאבתם

Josh 5:6 refers back to the nation "who had not listened to YHWH's voice, to whom YHWH had sworn that He would not have them see the land that YHWH had sworn to their fathers . . ."

אשר לא שמעו בקול ה׳ אשר נשבע ה׳ להם
לבלתי הראותם את הארץ אשר נשבע ה׳ לאבתם.

The allusion is direct.

I also noted that there is an allusion to J's Hamor of Shechem in the Abimelek story in Judges (9:28) and that there is an allusion to Abimelek in the Uriah story in the Court History (2 Sam 11:21).

There are a number of other cases. References to the Red Sea event in Joshua allude explicitly to the J version as well as to the Song of the Sea, which is embedded in J, and on which J depends, as Baruch Halpern has shown.[6] For example, Rahab says: "everyone living in the land is dissolving before you" (נמגו כל ישבי הארץ מפניכם Josh 2:9, 24). The Song of the Sea says: "everyone living in Canaan is dissolving (נמגו כל ישבי כנען Exod 15:15). (Other allusions: Josh 3:14, cf. Exod 15:8, 16; Josh 3:17, cf. Exod 14:21; Josh 4:23; 5:1, cf. Exod 15:16.)

In Judges (11:15–22), Jephthah refers explicitly to Moses' having sent messengers to Edom (and Moab) in Numbers (20:14–21; 21:12f., 21–26), which in turn plays upon Jacob's having sent messengers to Esau, who is Edom, in Genesis (32:4). Indeed, the words "and he sent messengers" (וישלח מלאכים) occur seven times in Kings, four times in the Chronicler's work, and once in E, but twenty-four times in this collection.

The narrative of David's and Saul's relations with the Gibeonites in the Court History (2 Samuel 21) relates back to the oath sworn to the Gibeonites by Israel, which is found in this collection in the book of Joshua (chap. 9). And the story of Israel's oath to the Gibeonites in Joshua, in turn, appears to be dependent on the command in the J Decalogue against making a covenant with the residents of Canaan (Exod 34:15).

The point of these signs of narrative continuity and allusion: this is a *connected* group of texts. The same group of texts that have the common and frequently unique body of terminology also tells a continuous, related story. The merging of these two characteristics—common language and connected continuity—indicates the unity of this corpus.

IV. Similarity of Whole Accounts

Another line of evidence is the similarity of entire sections within this collection. In the Introduction I referred to the cluster of similar language and images in the Amnon-and-Tamar and the Dinah-and-Shechem accounts. There were also the readily visible similarities of the story of Lot in Genesis 19 and of the cut-up concubine in Judges 19. And I detailed the cluster of common themes, language, images, and details in the succession narratives of Jacob and of David.

In addition to these, there are the similarly described accounts of lotteries in the portrayal of the identification of Achan in Joshua 7 (16–18) and of the designation of Saul in Samuel B (1 Sam 10:20f). The account in Joshua begins: "and he brought Israel forward by its tribes."

ויקרב את ישראל לשבטיו

The account in 1 Samuel begins: "And he brought all of Israel's tribes forward."

ויקרב את כל שבטי ישראל

Joshua continues: "and the tribe of Judah was taken by lot."

וילכד שבט יהודה

Samuel continues: "and the tribe of Benjamin was taken by lot."

וילכד שבט בנימין

And so on. We might explain this by saying the common wording is simply a natural result of the stories' being both about lotteries. But, in the first place, there are plenty of ways to describe lotteries. Indeed, there are descriptions of lotteries in the books of Jonah and Esther, and they do not have the same wording as these two accounts. And, again, the point is the *density* of the similar language here. They are too close for coincidence or common themes to explain. We might argue that their similarity is the result of the fact that they are both part of the Deuteronomistic history, the great work that extends from Deuteronomy to Kings, and that the editor of that seven-book work deliberately *made* them similar. But this answer leaves us hanging with two questions: (1) Why would he do that?! (2) Why does it happen so much in these chapters but not in the *rest* of the Deuteronomistic history? The fact is that these clusters of familiar words and phrases form an interlocking pattern that you will see all through the texts that I have singled out, but not through the other texts that surround them.

There is a less obvious but still visible cluster of similarities of language and images between the story of Jacob's first revelation at Beth-El in J and Samuel's at Shiloh in Samuel B:

Both, of course, are about revelation and sleep. Jacob has just left his father, who has been described as weak-eyed (Gen 27:1):

"his eyes were too dim for seeing."

ותכהין עיניו מראות

Samuel is with Eli, who is described as weak-eyed (1 Sam 3:2) with the same term (a term that occurs only in this corpus of texts):

"his eyes had begun to be dim; he was not able to see."

ועיניו החלו כהות לא יוכל לראות

In J, YHWH "stands over" Jacob: **והנה ה׳ נצב עליו** (Gen 28:13); in 1 Samuel, a corresponding term is used for the deity's presence: **ויתיצב** (1 Sam 3:10).

In both accounts the hero "lies down": **שכב** (Gen 28:13; 1 Sam 3:2, 3, 5, 6, 9).

Samuel is "in his place": **במקומו** (3:9); Jacob arrives "in the place": **ויפגע במקום** (28:11).

1 Samuel begins with the notice: **אין חזון נפרץ** (3:1); God tells Jacob: **ופרצת** (28:14).

There are thus at least five parallels of dense clusters of terminology spread through this connecting body of texts, occurring in J, Joshua, Judges, 1 Samuel, and 2 Samuel. In the past, we have explained such parallels as editorial reconciliation, or as one author imitating another, or as one author being influenced by the actual history reflected in the other story, or as both authors deploying common formulae from oral traditions. But none of these explanations comes to terms with the converging lines of evidence, viz., with the fact that these parallel clusters fall in a connected body of texts, which are pervaded with characteristic terminology, and which allude to one another and depend upon one another. Moreover, as I set out to demonstrate in the Introduction, there is a literary relationship among these works. It is not just that there is a large set of parallel words in each of these pairs of stories. Rather, the content of the stories is part of a visible structure. As my colleague David Noel Freedman particularly demonstrated in the case of the Dinah-and-Shechem and Amnon-and-Tamar episodes, the two stories are necessarily linked, so that they produce an "echo effect." He concluded that "we can recognize that not only is a single mind at work but that the stories belong to a single literary endeavor."[7]

V. Repeated Prose Images

Another body of evidence derives from the presence of numerous cases of repeated prose images. A scene that is pictured in one story recurs in another part of the collection, along with some recurring word or phrase to confirm that the parallel pictures are related. It is not just that the J account of Moses at the burning bush (Exod 3:5) and the account of Joshua's encounter at Jericho (Josh 5:15) both include the instruction, "Take off your shoes from your feet, for the place on which you are standing: it's holy ground." This in itself proves nothing, because we can imagine two writers drawing on a famous line, or one writer imitating another. But there are enough of such repeated images, flagged by common terminology, that fall within this group as to constitute evidence that these are part of a design. Thus:

The J story of Shechem (Gen 34:24f.) and the Joshua story of Gilgal (Josh 5:8) both depict cases of men sitting and healing after circumcision. And both refer to the absence of circumcision as "disgrace"—*ḥerpâ* (Gen 34:14; Josh 5:9).

Both Samson in the book of Judges (13:5; 16:17) and Samuel (1 Sam 1:11) in Samuel B receive the instruction, "and a razor shall not go up on his head."

Both the J story of the three visitors to Abraham in Genesis (18:5) and the story of the concubine from Bethlehem in Judges (19:5) refer to "satisfying one's heart" with "a bit of bread." The expression does not occur anywhere else in the Hebrew Bible.

Likewise: the J version of the story of Joseph pictures his brothers reporting that the powerful man they met in Egypt (who was in fact in Joseph) told them, "You won't see my face unless" they bring their brother Benjamin to him (Gen 43:5). And then, in the Court History, David tells the Israelite general, Abner, "You won't see my face unless" he brings David's wife Michal (who happens to come from the tribe of Benjamin) to him (2 Sam 3:13).

Both the J story of Abraham's oath to his servant (Gen 24:8, 41) and the Joshua story of the spies' oath to Rahab (Josh 2:17, 19, 20) offer the option of voiding the oath under certain conditions. In both cases, the expression is: to be "free of the oath" (נקי משבעה).

Both Rahab in Joshua (Josh 2:15) and Michal in the Court History (1 Sam 19:12) are pictured as saving men in the same manner and in the same wording: "and she let . . . down through the window" (ותוריד ... בעד החלון).

Both Rahab in Joshua (Josh 2:21) and Tamar's midwife in J (Gen 38:28) are pictured as tying a red cord, and in similar wording. Tamar's midwife: ותקשר על ידו שני. Rahab: ותקשר את תקות השני.

In another cluster of language and images, the story of the Gibeonites' deception in Joshua includes two rare biblical terms in a single verse: טלוא (meaning "spotted") and נקוד (meaning "streaked"). These terms are used in Josh 9:5 to describe the Gibeonites' worn clothing and provisions. But these are the same terms that are used together repeatedly in the J story of Jacob's deception of Laban in Genesis 30 (vv. 32, 33, 35, 39). Jacob gets all the sheep that are spotted or streaked. These two terms do not occur together anywhere else in the Bible. And a few verses later, Joshua says to the Gibeonites, "Why did you deceive us?"—which is what Jacob says to Laban back in Genesis (29:25).

Several times in this collection, there is pictured a case of two wives, one of whom is barren at first but who later gives birth: Sarah and Hagar, Rachel and Leah, Hannah and Peninnah.

Several times in this collection there is an account of a man being buried in his father's tomb. This is the fate of Jerubbaal and also of Samson in the book of Judges (Judg 8:32; 16:31). It is the fate of Asahel and of Ahitophel and of Saul and Jonathan in the Court History (2 Sam 2:32; 17:23; 21:14). The words "He was buried in his father's tomb" do not occur anywhere else in the Hebrew Bible.

There is also a parallel in two scenes of the Israelites responding to their leaders: Moses at the Red Sea and Samuel at the inauguration of Saul. Moses says (Exod 14:13): "Stand still and see YHWH's salvation that He'll do for you today."

התיצבו וראו את ישועת ה׳ אשר יעשה לכם היום

Samuel says (1 Sam 12:16): "Stand still and see this big thing that YHWH is doing."

התיצבו וראו את הדבר הגדול הזה אשר ה׳ עשה ...

In response to Moses: "the people feared YHWH, and they trusted in YHWH and in Moses His servant" (Exod 14:31).

וייראו העם את ה׳ ויאמנו בה׳ ובמשה עבדו

In response to Samuel: "all the people feared YHWH and Samuel very much" (1 Sam 12:18).

וירא כל העם מאד את ה׳ ואת שמואל

The issue here is more than similarity of language. It is the redeployment of prose images. The parallels in the language serve to indicate that these similar images were consciously selected. They are not just cases of an old image known from oral narrative being used in two different places. They are part of a designed construction.

As proof of this, look at this parallel case: the renaming of Beth-El and of Dan.

> And he called that place's name Beth-El, though in fact Luz was the name of the city at first.
>
> (Gen 28:19)

> And they called the city's name Dan . . . , though in fact Laish was the name of the city at first.
>
> (Judg 18:29)

The wording of the two notices is identical: *we'ûlām . . . šem hā'îr lāri'šōnâ.* And this phrase occurs nowhere else in the Hebrew Bible. But note also that this is not a chance phraseological connection of just any two cities. This phrase is applied specifically to *Dan* and *Beth-El,* the two cities that mark the northern and southern border of the kingdom of Israel, the two cities that are said to house Jeroboam's two golden calves. And we cannot claim that a mysterious (Deuteronomistic?) editor inserted these two notices, because there already was a notice of the Luz–to–Beth-El change in the E source (Gen 35:6–7)—and we do not find biblical editors deliberately creating doublets. (By any reckoning, the Deuteronomistic historian did not get hold of this text until after J and E had already been combined.) The doublet in J and E indicates that the report was native to both. (And note that the E report uses different wording.) What we have here is parallel images, parallel terms, and a connection between the two parallel cases that indicates that they were designed to go together.

VI. Similarity of Technique

Another line of evidence further supports the perception of a conscious process in the relationship of these texts. It is the evidence of technique. First, the art of punning—paronomasia—is employed in the fabric of most of the texts in this collection. J is rich in puns of various sorts, including alliterative puns, such as in the oracle to Rebekah (Gen 25:23–24):

לאם מלאם...וימלאו

Or puns related to story content, such as the use of the root *yṣr* ("to fashion") for creation three times in J (Gen 2:7, 8, 19) followed by the reference to "every inclination *(yēṣer)* of their heart's thoughts," which introduces the flood story (Gen 6:5). Or puns upon names in the story, such as the river Gihon, which flows out of Eden, where later a snake is told, "you'll go on your belly *(geḥōnekā)*." Or puns of allusion, as indicated by the presence of six occurrences of the root *rḥb,* of the name Rehoboam, the first king of the divided southern kingdom, while the root never occurs in E.

Such forms of paronomasia occur in other texts from this collection as well. In Judges, when Abimelek is at *Shechem* he carries on his shoulder: *šikmô* (Judg 9:48). And Israel's enemies are pictured as troublesome alliteratively: וירעצו וירצצו

In 1 Samuel, there are the well-known plays on the name Saul in the course of the birth story of Samuel: הוא שאול לה׳ (1 Sam 1:27, 28). This is commonly thought to be an appropriation of a Saul birth story for Samuel, but that need not be the case. The proleptic reference to Saul is simply a good pun for an audience who knew Saul was coming, just as they knew of a Rehoboam. The absence of a *mem* in the root that is used in the etiology of Samuel's name presents no problem when we consider that in this very collection of texts we find Noah's name coming from the words זה ינחמנו (Gen 5:29). Puns need not be based on etymology (as וירעצו וירצצו obviously is not).

Likewise in the Court History, we find such puns as this. In the wake of the Absalom rebellion, David is described as "coming in peace"—Hebrew *b' bšlwm,* playing on the letters of the name Absalom: *'bšlwm* (2 Sam 19:25). And this comes in the same chapter as the punning metathesis: ויבא יואב ("And Joab came"—19:6). This is the work in which David's liaison with Bathsheba results from his seeing her as she washes, and then when he wants Uriah to sleep with her he tells Uriah to go home and wash his feet. And this is the work in which Abigail explicitly plays upon her husband's name, Nabal (meaning "fool"), calling him a fool (1 Sam 25:25). And at the time she says this she is bringing bottles of wine (25:18), and the word for "bottle" is Hebrew *nēbel.*

Now, it is true that puns are a device in many prose and poetic works in the Bible, but we have to take account of them in the evidence for three reasons. First, as a general observation, the presence of puns of various sorts in so many of the texts of this collection is one more element of commonality among them. Second, the puns sometimes serve to bind the work. Thus, for example, the work starts with the pun on the name of the river Gihon, and at its conclusion it refers three times to Solomon's becoming king at the Gihon (1 Kings 1:33, 38, 45). Third, *the puns play on some of the very terms that we observed in the first line of evidence as recurring in this collection: nabal, raḥaṣ.* That is important because it is further evidence that the author *consciously* deployed and cultivated these terms in these texts. Thus, recall that the term for the dimming of an old man's eyes (Hebrew root *khh*) occurs three times across this collection. Then note that in the third occurrence, the notation that Eli's eye had dimmed *(khh)* is followed by the deity's criticism of Eli for not "restraining" his sons from crimes they committed (1 Sam 3:13). The word for "restraining" is the Hebrew root *khh,* punning the dimming of his eyes. Punning on the very terms that recur in this collection reveals an author who was thinking about these words, deploying them with literary purpose. Their disproportionate occurrence in this collection is not just a matter of random distribution.

There are other matters of technique in this collection as well. J is rich in irony, as I discussed in the Introduction. Jacob appropriates his brother's birthright, and then he is cheated out of Rachel because, Laban says, it is not the custom to give the younger daughter before the firstborn. Jacob deceives his father with his brother's clothing and the skins of a goat, and then Jacob's own sons deceive him about their sale of Joseph with their brother's clothing dipped in the blood of a goat. And so on—the cycle is rich in a trail of ironies.

Likewise in the Court History, David sees Uriah's wife, Bathsheba, from the roof, and he takes her; and later Absalom takes David's concubines on a roof—and later still, a watchman on a roof sees the runners coming with the report of Absalom's death. Likewise in Judges, when Eli does not restrain *(khh)* his sons from wrongdoing, and later we find that his eyes had begun to weaken *(khh),* this is more than just a pun. It is an expression of irony.

And, again, the cases of irony are several times constructed with the very terms that recur in this collection of texts (עזים, גג, כהה), which reinforces our perception of conscious deployment of these words.

In considering technique, I also noted that there is an uncommon degree of character development in this collection of texts and that all three uses of parable in the Hebrew Bible fall in this collection (Judg 9:8–15; 2 Sam 12:1–4; 14:5–7).

Common elements of technique thus converge with the other lines of evidence that there is a literary relationship among this group of texts.

VII. Theme

Thematic evidence converges as well. We have already observed that a blatant concern with succession is developed in both J and the Court History. We have also seen that both elaborately develop the theme of deception and corresponding recompense, especially within families. Both develop family relations, especially hostile relations, particularly between fathers and sons, but also between mothers and sons, fathers and daughters, brothers and sisters, brothers and brothers, cousins, uncles and nephews, and husbands and wives. It is by no means a given that such matters should figure in narrative; and, in fact, in most of the rest of biblical narrative they do not figure at all. For example, the other narrative works in the Torah do not develop the relations within Moses' immediate family; only J tells a story involving his mother and sister, and, for that matter, only J has a story in which his wife acts.[8] And we have observed that the more limited familial theme of fratricide is powerfully developed in this collection, with at least seven cases from Abel to Adonijah.

We have also observed that the collection develops the matter of drunkenness five times, and in every case the drunken person is deceived. And we have observed that the collection weaves in nine spy stories. And we have observed an array of stories involving sexual matters in numbers without a remote parallel in any other biblical work—in fact, there are far more of such episodes here than in all the rest of biblical prose put together. We have also observed that this collection repeatedly depicts the dependent position of women functioning in patriarchal society, starting with the etiological declaration to the first woman: "your desire will be for your man, and he'll dominate you," and proceeding to the cases of Sarah and Hagar, Rebekah, Jacob's wives and their maids, the concubine in Judges, Dinah, Tamar (Judah's daughter-in-law), Jephthah's daughter, Hannah, Samson's mother, Abigail, Michal, Bathsheba, Tamar (Amnon's half sister), and Abishag.

Also, there is a continuing thematic concern with Judah and its interests. In addition to the Judean concerns in J that I noted earlier, there are other signs of J's disproportionate interest in Judah. In J, it is Judah who saves Joseph from his brothers, it is Judah who offers surety to Jacob of Benjamin's safe return from Egypt, and Judah who gets the birthright (or preeminence). J includes stories of the elimination of

Reuben, Simeon, and Levi from the running for the birthright. J offers the additional story of Judah and Tamar, which culminates in the birth of the eponymous ancestors of the Judean clans: Perez and Zerah. Moreover, it favors Perez, the clan from which David and the royal house came: thus it adds a story of Perez's supplanting Zerah and thus becoming the firstborn (Gen 38:27–30). And it further diminishes the clan of Zerah in the story of Israel's defeat at Ai in the book of Joshua: Israel suffers in that story because Achan, *of the clan of Zerah,* violates the law of complete destruction *(ḥerem)* at Jericho (Josh 7:1, 17, 24).[9] In J, only Caleb—of Judah—is a faithful spy. The Court History likewise develops the ascendancy of the Judean monarchy, the defeat of attempts of the northern tribes of Israel to break with Judah, and a continuing interest in Caleb's family. Also in the Court History, Judah is the first to bring David back after Absalom is defeated.

Meanwhile, this collection has more than one indicator of hostility toward Shechem, the first capital of the northern kingdom. According to E, Israel's ancestors bought it; according to J, they massacred it. And Shechem is again dishonored in Judges 9. This work is continuously a work of Judah.

In this regard, we should take note that many biblical scholars have observed connections between J and what they took to be the *history* of Judah, but they were in fact observing parts of the chain of connections in this group of texts. For example, Gary Rendsburg observed an intriguing set of parallels between the J story of Judah and Tamar and the Court History of David. He concluded that there was a connection of Genesis 38 to "the life of David." I suggest that we do not know the life of David. We know the literary work known as the Court History. This methodological point applies not just to Rendsburg but to a number of treatments—such as those of Ronald Clements and Walter Brueggemann—that relate J matters in Genesis to King

David, when their source for David is specifically the Court History. Clements says, "It is impossible to avoid the conclusion that the Yahwist himself saw an important connection between Abraham and David." Brueggemann says, "The J construction in Genesis 2–11 is dependent upon the career of the sons of David in their quest for the throne." And he adds, "The Genesis stories may be for the Yahwist another way of writing about the dynasty . . ."[10] What they identify as the *historical* background of a text is rather a close *literary* relationship between two texts. Their sensitivities to connections were on the mark, but the connections are not between J and the history of Judah; they are between the Judean theme in J and the Judean theme elsewhere in this literary collection.

The matter of theme thus provides one more collection of evidence that aligns consistently with the other corpora of evidence that this work is a single, united literary construction.[11]

The evidence that these are a related, united body of texts thus converges from a variety of areas: terminology, continuity, allusion/quotation, similarity of entire sections, repeated prose images, technique, and recurring theme. The linguistic evidence supports their origination from the same period. And the literary structures that I began to describe in the Introduction and the Afterword further support this evidence that they are a single literary work.

In the Introduction, I discussed the sense that literary critics (Auerbach and Saville) had that J and the Court History came from the same author. Numerous biblical scholars, as well, have had the intuition that these works were related. Steuernagel had two Samuel sources, one of which was likely to be from a "Yahwistic school."[12] Budde had J and E sources that extended to the Court History as well.[13] But the chain of connections that bind J and the Court History, as well as the associated texts that come between them, was not observed, and so these views were generally abandoned in scholarship. Also contributing to their demise—in fact, the main factor—was the wide acceptance of Martin Noth's view of the Deuteronomistic history as a work that made a clean break between the Tetrateuch and the books that followed.[14] But imagining a complete separation of the Tetrateuchal sources from what followed was an enormous oversimplification of what Noth's view entails. Martin Noth was careful to make clear that the Deuteronomistic historian did not write all the history himself but rather used written sources which he incorporated into the work. In the case of the narrative of David's court, Noth wrote: ". . . the extant version of the story of King David begins by following the remains of the traditional account of the rise of David and is subsequently based on the story of the succession. Here Dtr intervenes only occasionally and even then contributes very little."[15] Recognizing the existence of the Deuteronomistic history (as I do[16]) does not release one from analyzing the materials and process that produced that history. It was not composed out of whole cloth by a single historian sometime after the fall of Judah in 587 B.C.E.[17] On the contrary, the Deuteronomistic historian explicitly acknowledges that he used sources. And the evidence here indicates that the work that we have identified was the largest of his sources. Recall, for example, that the recurring words and phrases that we observed do not match the boundaries of the full Deuteronomistic history (i.e., from Deuteronomy to 2 Kings); they match the boundaries of the work that we have identified. They go up to 1 Kings 2 and then stop abruptly. The other categories of evidence converge with this terminological pattern.

The scholarship of the last century on the editing of the Bible's main narrative works thus had the ironic effect of leading us away from the trail that earlier biblical scholars had begun to follow. The work that we are observing here had the strange fate to fall into the hands of several different editors. R (and RJE) drew on it as a source for the Torah. They used only the first part of it, and they handled it according to the standards, needs, and methods that characterized their formation of the work that became the Pentateuch. The Deuteronomistic historian drew on it as a source for his history. He used the latter part of it, and he handled it according to his respective method. The result of this dual editorial process was to make it even harder to identify and associate the parts of this work.

There are many implications if this analysis is correct:

1. Insofar as it shows a body of text that has linguistic affinities to J but not to E, it is a new and strong verification of the identification of J in the Pentateuch. It is likewise a new verification of the existence of J and E as two separate works.

2. It is also a reconfirmation of the distinction of Samuel A and Samuel B as the two sources of the book of 1 Samuel.

3. It, too, is a further confirmation of the separation of Judges 9–21 from chapters 1–8, a division that Baruch Halpern had arrived at on entirely different grounds.[18]

4. This is also significant as evidence that at least one of the Pentateuchal sources extends into Joshua (and beyond).

5. At the same time, this does not contradict the view that a united work, the Deuteronomistic history, extends from Deuteronomy to Kings. On the contrary, it provides useful information on how the Deuteronomistic historian used sources. Some scholars have spoken of the Deuteronomistic history as if its historian had written the entire work in his own words, but that is not the case. The identification of the work of the first great prose writer enables us to see one of the sources that the Deuteronomistic historian used in assembling his record of Israel's past.

6. Moreover, this identification gives us a rare opportunity to compare how two different editors used the same source: RJE edited

the portion of this work that is now contained in Genesis through Numbers. The Deuteronomistic historian edited it as it appears in Deuteronomy through Kings. The most significant revelation in this regard is the deeply disappointing fact that the editor who combined J with E (known as RJE) was more willing to cut his sources than was the Deuteronomistic historian. Thus, when we read the material that is contained in the books of Joshua through 1 Kings 2, we find that it flows as an almost complete story, but when we read J in the Torah alone it has numerous breaks where portions were apparently removed.

These are among the implications if we are correct in identifying this as a united work. There are further implications if this work also was composed in the period of the divided kingdoms of Israel and Judah. The antiquity of the work is the subject of the next two sections.

2
THE ANTIQUITY OF THE WORK

My purpose is not only to establish that this is the work of a single great writer but that it is the work of the *first* great writer.

The Documentary Hypothesis has been a dominating model in biblical scholarship since the nineteenth century. The basic paradigm has become so firmly established and widely accepted that much of the present generation of scholars' time has been spent in seeking some arena in which we could make new contributions. Some scholars have concentrated on the artistry of the final product, the so-called "text as it is," usually minimizing the significance of the source works and of the authors who wrote them. And those scholars who continue to specialize in author-centered studies of literary history—in other words, those who still work on the sources—have had to find their arena as well.

The last model-changing contribution to this literary history is usually attributed to Martin Noth, who established in the 1940s that the books of Deuteronomy through Kings were composed as a single continuous work, called the Deuteronomistic history. The four biblical books that precede the Deuteronomistic history (Genesis through Numbers, also known as the Tetrateuch) were a separate work which, according to Noth, was joined to the Deuteronomistic history by a subsequent editor. After Noth, if one wanted to make a comparably major contribution, the only realistic options were either to argue that the Priestly corpus (P) was composed early, which is to say prior to the exile of the Jews to Babylon in 587 B.C.E., as the Israeli scholar Yehezkel Kaufmann had argued against the German scholar Julius Wellhausen from the beginning—and which remains the primary dividing point between most Jewish and Christian scholars in an otherwise fairly ecumenical field. Or one could argue that J and/or E were late, which is to say exilic or post-exilic. Lately, some scholars have raised arguments for the late composition of J, and I think that one of the reasons why these arguments have received as much prominence as they have is that they constitute a substantial modification of the standard model, not just a reshuffling of the cards. And, theoretically, it is healthy and refreshing for the field to challenge the most basic components of the structure periodically. I have no problem with challenging the standard model, obviously, because I am one of those who are persuaded that the Priestly narrative is pre-exilic, as I have argued in earlier books.[19] But it seems to me that the strongest of the new evidence that has come out in this generation adds further support to the already strong case that had been worked out in the last century for the very early composition of J.

Perhaps the most important aspect of this evidence is its strength as a *cumulative* case. After all, the widespread acceptance of critical scholarship in the first place owed precisely to its cumulative strength and not to any one type of evidence. This is demonstrated and emphasized in my entry on "Torah (Pentateuch)" in *The Anchor Bible Dictionary,* which was meant to be the largest collection of evidence for the basic hypothesis that has thus far been made in one place. I emphasize this point—the cumulative case—many times there because I think that it has not been sufficiently recognized, either in scholarly or in popular treatments. It was not that there were doublets. Everyone had been aware of that for two millennia. And it was not that there were contradictions in the story, and it was not that there were different names of God. It was that the doublets could be divided into groups, *and* the different names of God fell consistently into one of these groups or another, *and* this division resolved most of the contradictions. These and other categories of evidence accumulated, so that many different bodies of data were all pointing in the same direction. Traditional rabbinic and Christian scholarship had offered explanations of the doublets, contradictions, and so on, all along, but it was doing it one verse at a time. If there were two thousand such problems, there were two thousand separate explanations for them. Critical scholarship explained it all with vastly fewer premises. The success of critical scholarship was a quintessential demonstration of the compelling quality of Occam's razor.

I've reviewed this history because I think that the evidence concerning the identification of J, as well as E, which was accumulated from the seventeenth to the early twentieth century, is still compelling, and the recent evidence compiled by others and myself has further confirmed it. I am so concerned about the works in recent biblical scholarship by colleagues who claim to have proven that J was written late because these colleagues have never recognized the cumulative weight of all of this evidence. Indeed, they have sometimes failed even to mention it, much less to respond to it. These are several categories of evidence, containing several thousand individual points of data that are consistently explained within the theory. Often, a scholar proposing a new model has begun by challenging some of the individual points of data in the theory. The scholar might argue, for example, that a city mentioned in J was built later than the time in which J was supposed to have been written in the theory. But such points of data constitute a case for refinement, which is a basic process in the formulation of any theory in any field. They are grounds to reconsider the evidence for the dating of the city in the Bible or in the archaeological or literary sources outside of the Bible. They are grounds to consider whether scribes or editors made changes to the names of places and persons after the original composition of a work. But, weighed against the thousands of points of data that still consistently fit the theory in converging lines of evidence, they are not grounds for throwing out the theory and accepting the scholar's new proposal instead.

One of the reasons why this matters so much is that these converging lines of evidence mean that these texts are the first known lengthy works of prose ever composed on earth.

The case is as follows:

First, there are both a J and an E. I mean two originally complete texts, not a J source with a group of additions. Critical scholars agree to a large extent on the identification of Priestly and Deuteronomic texts. But the strongest doubts have been about the identification, or even the existence, of E. Texts that we used to think bore obvious signs of being E are now claimed to be part of J or to be small, independent stories. This is not a small point. If J and E sources are preserved along the lines that I am identifying here,[20] then the models for a late J must be wrong, as we shall see. What is the evidence for the existence of both J and E:

First, there are doublets within the material in question, including: the Genesis wife/sister stories (Gen 12:10–20 and 20:1–18); the naming of Beer-sheba (Gen 21:22–31 and 26:15–33); the story of selling Joseph into slavery: what the original plan was, which brother figured prominently, and who sold him (Gen 37:2b, 3b, 5–11, 19–20, 23, 25b–27, 28b, 31–35; 39:1 and 37:3a, 4, 12–18, 21–22, 24, 25a, 28a, 29–30); the etymology of the name of Moses' son Gershom (Exod 2:22; 18:3); the burning bush (Exod 3:2–4a, 5, 7–8 and 3:1, 4b, 6, 9–15); and the theophany at Sinai (Exod 19:2b–9, 16b–17, 19; 20:18–21 and 19:10–16a, 18, 20–25).[21]

Second, there are contradictions within the material: for example, the matter of whether Joseph is sold by Midianites or by Ishmaelites is utterly irreconcilable, as is well known to anyone who has ever tried his or her hand at resolving it (Gen 37:25–36; 39:1); plus the matter of Moses' father-in-law being first Reuel then Jethro (Exod 2:16–18; 3:1; 4:18; 18:1); or the mountain being first Sinai then Horeb.

Third, terminology: The best-known case is the divine name. Though periodically challenged in scholarship, this remains a strong indication of authorship. J consistently excludes the word *'ĕlōhîm* (God) *in narration.* Individual persons in J use the word *'ĕlōhîm,* but the narrator of the story in J does not. Instead the narrator consistently calls the deity by the proper name YHWH, with perhaps one exception out of all the occurrences in the Pentateuch.

E, however, maintains a distinction: the name YHWH is not revealed until the time of Moses (Exodus 3). Prior to Moses in E the deity is referred to simply as God *('ĕlōhîm)* or by the name *'ēl,* but not by the name YHWH. There are two or three possible exceptions out of all the occurrences in E.

(P maintains the same distinction as E, that the divine name is not known until Moses (Exod 6:2–3), with one possible exception out of hundreds of occurrences.[22])

The consistency of this phenomenon through well over two thousand occurrences makes it a very serious point. One might think that this is merely circular because a scholar can simply identify any bit of text as J, E, or P according to the name of God. But such willy-nilly division of verses by divine names should break down. There should frequently be passages that are thus identified as J because they include the divine name but which then show other indisputable signs of being P or E. *But that does not happen.* Rather, when we divide the divine names by these very fixed and almost exceptionless criteria, we still arrive at consistent source divisions. And within those divisions, all doublets separate and line up consistently, and numerous contradictions resolve.

Thus, for example, when we separate the doublet of the name Gershom, we find that one account of the etymology occurs in a passage that identifies God in narration as *'ĕlōhîm* (Exodus 18), while the other account of the same etymology occurs in a passage that does not (Exodus 4). And the passage that identifies God as *'ĕlōhîm* in narration identifies Moses' father-in-law as Jethro, as in other E texts; while the passage that does not use *'ĕlōhîm* identifies the father-in-law as Reuel. Such examples of consistent, mutually supportive lines of evidence are now so numerous as to form a powerful case for the identification of two definite sources, J and E.

Fourth, there are consistent characteristics within each source text. For example, the Tabernacle figures in E (three times) but never in J.

Fifth, narrative flow: It is sometimes possible to identify and separate two stories, one J and one E, that have been merged, and one can then read each as a continuous, flowing narrative. And these continuous narratives remain compatible with the evidence of doublets, terminology, contradictions, and consistent characteristics. Examples of such parallel, complete, continuous narratives are Jacob at Beth-El (Genesis 28); the sale of Joseph (Genesis 37); and the burning bush (Exodus 3).

The identification of the lengthy prose work that extends from J to the Court History now confirms this division of J and E. We have seen a group of texts that one can isolate in the books of Joshua through 1 Kings 2 that have a clear and striking connection to the texts I have identified as J, but no such connection to E (or to any other texts in biblical narrative). Whatever else we may make of the work I have presented and translated here, the very minimum point of it is: there are texts in the Bible that have a specific connection to the texts identified as J but not to those identified as E. This is a striking confirmation of all the other evidence for this separation into two bodies of text: J and E.

I've stated the case for there being both J and E because this separation and identification of the two classic sources is essential to figuring out when and where these works were composed. Once we have identified J and E, the historical referents of these two source texts lie in the period when the country was divided into the two kingdoms of Israel and Judah. **The historical referents of J disproportionately relate to Judah, and those of E disproportionately relate to Israel.** They had to be written while both kingdoms were still standing, before the kingdom of Israel was defeated by Assyria and ceased to exist in 722 B.C.E.

Elements of J that indicate a Judean provenance are:

In J Abraham resides in Hebron/Mamre (Gen 13:18; 18:1), the capital of Judah (and home city of Zadok, the Judean high priest of David and Solomon).

Meanwhile, Shechem was the capital of Israel, built by Jeroboam I, the king who had rebelled against Judah; and the J account of Israel's acquisition of Shechem is derogatory, involving the taking of Dinah by Shechem and the massacre of the city by Simeon and Levi (Genesis 34). In J: Hebron is good, and Shechem is bad.

The J birth accounts of the eponymous ancestors of the tribes include only: Reuben, Simeon, Levi, and Judah (Gen 29:32–35). Only Judah, among these tribes, existed with a territorial identity in the era of the monarchy. Moreover, J includes the story of Reuben's taking Jacob's concubine and the story of Simeon's and Levi's massacre of Shechem. As confirmed in Jacob's deathbed blessing in Genesis 49, a poetic text included in J, these acts result in the preeminence passing to the fourth son: Judah (49:3–12).

In the J story of Joseph, Judah is the brother who saves Joseph from the other brothers' plans to kill him (Gen 37:26–27; 42:22). And it is Judah who promises Jacob that Benjamin will survive the journey to Egypt (Gen 43:8–9).

J alone includes a lengthy story from the life of Judah (Genesis 38), culminating in the birth of Perez, the eponymous ancestor of the clan from which the Judean royal family was traced.

In the J spies story, the scouts whom Moses sends see only Judah and miss the rest of Israel (Num 17–20, 22–24).

Also in the J spies story, the one favorable spy is Caleb. The Calebite territory was located in Judah and included Hebron.

J includes a lengthy account of the birth, youthful relations, and break between Jacob and Esau/Edom. These stories reflect the kinship and historical relations with Edom on several points. J also includes the list of the kings of Edom (Genesis 36). There are no equivalent stories or records in E. Judah bordered Edom; Israel did not. And Judah is claimed to have dominated Edom from David's reign until Jehoram's.

The iconography of J corresponds to the situation in Judah. (1) The ark was located in Judah. J includes a description of the ark's movements in the wilderness (Num 10:33–36), and J associates the presence of the ark with military success (Num 14:41–44); but the ark is never mentioned in E. (2) The J Decalogue only prohibits making *molten* gods (Exod 34:17). This prohibition denounces the golden calves of northern Israel, which were molten, without denigrating the golden cherubs of the Temple in Judah, which were wooden and gold-plated. (3) In J, cherubs are used to guard the path to the tree of life, also consistent with the cherub iconography of Judah. But cherubs are not mentioned in E.

Meanwhile, the elements of E that indicate a (northern) Israelite provenance are:

In E Jacob struggles with God and, etiologically, names the site of this event Peni-El (Gen 32:31). The Israelite city of Peni-El was built by Jeroboam I, the founding king of the northern Israelite kingdom (1 Kgs 12:25).

In E there is an account of Joseph's deathbed wish to be buried in his homeland and an account of the Israelites taking his remains during the exodus. The traditional site of Joseph's grave was at the city of Shechem (Josh 24:32), which was also built by Jeroboam and served as the capital of Israel for a time (1 Kgs 12:25).

In E, the territory around the city of Shechem is acquired by peaceful purchase (Gen 33:18–19) in contrast to J's story of its violent acquisition (Genesis 34).

The E birth accounts of the eponymous ancestors of the tribes include: Dan, Naphtali, Gad, Asher, Issachar, Zebulon, Ephraim, Manasseh, and Benjamin—all of the tribes of Israel, but not Judah

(Gen 30:5–24a; 35:16b–19).[23] Further, in E the birthright goes to Joseph, thus creating the (northern) tribes of Ephraim and Manasseh. And in E Ephraim is favored over Manasseh (Gen 48:13–20), corresponding to the historical preeminence of Ephraim (Jeroboam's tribe). The term for the additional portion thus awarded to Joseph is the unusual *šekem* (Shechem—48:22), punning on the name of the Israelite capital city.

In the E Joseph story, Reuben, instead of Judah, is the brother who saves Joseph from the other brothers' plans to kill him (Gen 37:21–22), and it is Reuben who promises Jacob that Benjamin will survive the journey to Egypt (Gen 42:37).

In E the Egyptian taskmasters are identified as "officers of corvée" (*śārê missîm*—Exod 1:11). The northern Israelite tribes' dissatisfaction with the Solomonic policy of *missîm* was an explicit ground for their secession, which, according to the report of 1 Kings, was initiated by the stoning of Rehoboam's officer of the *missîm* (1 Kgs 12:18).

The heroic role of Joshua is developed in E, but not in J. Joshua is depicted as being of northern Israelite origins, of the tribe of Ephraim.

Meanwhile, neither J nor E shows any awareness of the fall of the kingdom of Israel nor of the dispersion of the northern tribes. J's reference to Esau/Edom's breaking Israel's yoke from its shoulder (Gen 27:40) probably places its composition at least after Hadad's rebellion against Solomon or even after Edom's full independence from Judah in the reign of the Judean king Jehoram (849–842 B.C.E.). E offers few clues to narrow its composition further within the two-century period of division.

Possibly the strongest new evidence confirming the early composition of this literature is the linguistic evidence, including the data of Robert Polzin, Gary Rendsburg, Ziony Zevit, and Avi Hurvitz.[24] Hurvitz especially has sought to show that—philologically and linguistically—any attempt to "squeeze" so much of the biblical corpus into the narrow time span of the Persian/Hellenistic periods is an impossibility. We must listen to the linguists, for whom the linguistic-historical research of the Hebrew language demonstrates:

1. that Late Biblical Hebrew is an identifiable linguistic entity, which represents—lexicographically, grammatically, and syntactically—the post-exilic phase of the history of the Hebrew language;
2. that there is a clear "linguistic opposition" between Late Biblical Hebrew, on the one hand, and standard/classical Biblical Hebrew (BH)—employed in the texts that represent the pre-exilic period—on the other;
3. that this "linguistic opposition" between pre-exilic and post-exilic Hebrew, as manifested in our biblical corpus, is corroborated by a variety of extrabiblical sources, dated likewise to the pre- and post-exilic periods, respectively.[25]

From this evidence, J, E, and the Court History of David in 2 Samuel come from the earliest stage of Biblical Hebrew (BH), P's language comes from a later stage along this continuum, and all precede Ezekiel. *It is linguistically impossible for J to be exilic or post-exilic.*

On the basis of the evidence summarized here, there were both a J and an E, and both derived from the period of the divided kingdoms, prior to the Assyrian defeat of Israel in 722 B.C.E.

What is at stake here? The literary stakes are nothing less than the recognition and recovery of the work of the first writers of prose on earth. The linguistic stakes are the understanding of the development of the Hebrew language. The historical stakes are the understanding of the development of history writing, starting with the recognition that it began in Israel and not Greece. The literary-historical stakes are the establishment of the primary sources of both the redactor of the

Torah (identified as R) and the editor of the books of Joshua through Kings (the Deuteronomistic historian), opening the door to understanding better than ever the way in which the major biblical editors treated their sources. The literary-historical stakes also include the recognition of the work of RJE, the most ignored person in Pentateuchal criticism, who combined J and E so exquisitely that he has successfully fooled my colleagues who think that J and E are one work. This is an appeal to recognize the strength of the evidence of the last 150 years of scholarship: recognize first the authors and their works, and then their backgrounds will fall in place.

3
LATE FOR A VERY IMPORTANT DATE

In recent years some biblical scholars have argued that the stories that have been attributed to J and E are not early at all. They have suggested that these texts were composed hundreds of years later, in the exilic or post-exilic world. If they are right, then J and E are not the earliest works of prose literature, nor are they the earliest works of history. This view has become accepted among a large number of German scholars and has also acquired a following among some American scholars. A leading figure in Germany is Erhard Blum, following the work of his teacher, Rolf Rendtorff; in the United States the person most commonly associated with this view is John Van Seters.[26] At the 1995 meeting of the Society of Biblical Literature in Philadelphia a session was organized in which Professors Blum and Van Seters and I were invited to discuss, debate, and defend our understandings of these narratives. My main criticism of their positions that day and to the time of this writing is that, apart from my objections to their arguments for their late dates, they have not come to terms with the original mass of evidence for the antiquity of the texts. Van Seters has made his case in a series of books in which he attempts to form a picture of the Hebrew Bible's history that is, in his words, "in opposition to much of the prevailing view." In the first words on the jacket of his third book in the series (*Prologue to History*, 1992) his work is described more dramatically as a proposal "to turn Pentateuchal studies on its head." Quite aside from the question of whether their arguments for their own views are convincing or not, Van Seters and Blum are bound at least to address the original case that made this "the prevailing view" and the new evidence that has surfaced to support it in recent years.

As we have seen, the strongest argument for this view is not any single body of evidence but precisely the convergence of so many lines of evidence. In our meeting, Professors Blum and Van Seters did not so much as mention this cumulative argument in their presentations. When I pressed them on the point, Blum's response was, "But that is what we *all* do." He suggested that in all scholarship we try to mount as much consistent evidence as possible in support of our case. With due respect to a colleague whose intellect I admire, I have to say that that is not at all what we all do. All scholars may try to find as large a *quantity* of evidence as possible in support of their conclusions, but there is a difference between that and a constantly converging set of entire *lines* of evidence. Whole sets of doublets divide into groups, which consistently line up with regard to terms and names, the contradictions resolve, and the language of each group matches particular stages in the development of Hebrew. Such a consistent, regularly converging set of independent bodies of data is not what we all do. It is an extraordinary phenomenon that occurs rarely in the scholarship of any field. Those who argue for a late date of these works apparently do not yet appreciate the force of this phenomenon, much less respond to it.[27]

Likewise, a major development in the scholarship of the last two decades is the linguistic work by which we can observe the development of Biblical Hebrew, like all languages, through time. In work by scholars like Hurvitz, Polzin, Rendsburg, and Zevit, it appears that we can observe the stages in the development of Hebrew to which the various biblical prose works respectively belong. The results of this research are completely counter to Blum's and Van Seters's conclusions. The texts that they call late are written in the earliest, "classical," stage of biblical Hebrew: J and E come before the Deuteronomist. J, E, D, and P *all* come before Ezekiel. But Blum, Van Seters, and others do not take this on. Putting J in the exilic period without addressing the linguistic evidence is like putting Shakespeare in the twentieth century without addressing the fact that he certainly sounds different from the rest of us.

As for their own arguments, I believe the cases that scholars like Blum and Van Seters have made are flawed. I shall point out some of the flaws of fact and especially of method because I do not want to be accused of the same failure to address models that contradict one's own. My aim here is to note enough specific errors and general methodological missteps to reveal the weakness of these cases for moving these authors down toward the end of the biblical period.

I also want to keep in mind what is at stake. Some who criticize Higher Criticism and the Documentary Hypothesis have mentioned these recent works as evidence of its demise, but such works, to my mind, are not an attack on the documentary hypothesis. For the most part, they *are* the Documentary Hypothesis. Van Seters prefers to identify his view as a "supplementary hypothesis." That is, he argues that some of the texts were not written as independent works but rather were composed as additions to earlier texts. I am happy to accept that distinction. But my point remains that people who dislike source-critical biblical scholarship will not be any happier with this work than they are with any other work that sees J, P, and D texts in the Pentateuch. Van Seters and others are just dating the source J (and the E stories, which he considers to be part of J) to a much later time than most biblical scholars have dated it. When scholars attempt to place authors in the period that best suits the evidence, they are not challenging the hypothesis. They are *doing* the hypothesis: working it out, addressing the problems, seeking solutions. When they argue whether a text was composed as an independent document or as a supplement to an existing document, they are pursuing the *relationship* between the sources of the Torah. They are not differing on the essential fact of the existence of the sources. The significance of Van Seters's work on the hypothesis is that in his picture so very much of the Five Books of Moses was written so very late. In the second book of his series (*In Search of History,* 1983), he pictured the Deuteronomistic history (the books of Deuteronomy to Kings) as having been composed by a historian living after the Babylonian defeat of Judah and exile of a portion of its population in 587 B.C.E. In another earlier book (*Abraham in*

History and Tradition, 1975) and in *Prologue to History,* he attributed the texts that are usually known as J and E to someone he calls the Yahwist, who also lives after the exile, even later than the Deuteronomistic historian. (Van Seters excludes a few texts that he calls pre-Yahwistic.) And he informs us that he accepts the view that the remainder of the Torah, known in the hypothesis as P, was written latest of all, after the Deuteronomist and Yahwist.

That is, ancient Israel produced practically no history, no prose at all (at least none that survived), during all the centuries that it existed as a nation in its land. Late Jewish writers fashioned the history of their people after they had been evicted from their land. Also, according to Van Seters, even though these writers were living in the pocket of the Babylonians, they wrote under the influence of the historiographic traditions of the Greeks. (Blum differs with Van Seters on this point.) There is the question of how these late historians knew (or claimed to know) what had happened in all those earlier centuries. Van Seters allows for some available Mesopotamian sources, such as the Atrahasis flood account, for the Genesis stories, but he allows very few sources to the Deuteronomistic historian, who rather invented much of the history himself.

Van Seters's model has received approval from some scholars, especially in Germany, and his work has also been severely criticized by others, most recently and thoroughly by the American biblical historian Baruch Halpern.[28] At the 1994 meeting of the Society of Biblical Literature, held in San Francisco, Van Seters was offered an opportunity to defend his picture of the Deuteronomistic history. In the last question in that session the biblical historian J. Maxwell Miller asked Van Seters: If the Deuteronomistic historian was writing history by invention, and not from real historical sources, how did he manage to describe the actions of Pharaoh Shishak (in 1 Kgs 11:40; 14:25) so accurately? As Miller put it, how did he get Shishak in just the right "time and pew"? Van Seters responded that he assumed that there were monuments all over the country with information on them. Miller explained that this still would not enable a historian centuries later to locate Shishak so well in the right "time and pew" of an invented history, and so he asked again: If the biblical historian was just inventing, how could he possibly have gotten such details right? Van Seters's response was, "I wish I knew."

This troubles me because as I read Van Seters's third book in this ongoing project, I observed this same misunderstanding of method and this same failure to come to terms with evidence that flatly contradicts his model. For example, he simply mistakes the facts in the matter of the names of the deity. Recall that J consistently excludes the word *'ĕlōhîm* (God) *in narration.* Individual persons in J use the word "Elohim," but the narrator of the story in J does not. Instead the narrator consistently calls the deity by the proper name YHWH. As I said above, the consistency of the divine name distinction through well over two thousand occurrences makes it a very serious point. One must either accept it as reflecting different authors or one must make a strong attempt at explaining how this could possibly have happened. Van Seters, however, simply rejects the entire phenomenon on the grounds that "we have already seen a few instances where J has used Elohim alone instead of YHWH."[29] But the cases to which he refers are in quotation, not narration, and Van Seters has misunderstood the evidence. Even though J is nearly 100 percent consistent in discriminating between narration and quotation, Van Seters says, "the use of Elohim for YHWH is quite regular and indiscriminate in J."[30]

While simply misstating the evidence for the divine name distinction, Van Seters uses the subtlest twists of meaning in individual passages to argue against this distinction. In the story of Jacob's revelation and dream of the ladder at Beth-El (Gen 28:10–22), Van Seters eschews the blatant conflict in the divine names that has led other scholars to see this story as a combination of J and E, and he attacks the division on the following grounds:

1. In a passage attributed to J, Jacob is said to have "spent the night" (Hebrew *lwn;* 28:11a).
2. Then in a passage attributed to E, Jacob is said to have lain down (Hebrew *škb*) and dreamed (vv. 11b, 12).
3. Later, in another J passage, Jacob is said to have awakened (v. 16).

Van Seters insists that this reference to awakening *must* depend on the reference to Jacob's dreaming in the middle passage, so they cannot be from two separate sources. Why? Because, Van Seters claims, the word for "spent the night" *(lwn)* "does not include the notion of rest or sleep." Even if we accept Van Seters' narrow understanding of this root, what kind of argument is this? Can Van Seters really not imagine someone writing a story and saying "he spent the night in the forest" or "he stayed at their house that night" and later saying "he awoke the next morning"? Does the author have to refer explicitly to sleeping, or even to *dreaming,* for the reader to get it?

Blum, too, sometimes argues that the present form of passages constitutes a literary unity that is ruined by dividing these passages into originally separate J and E components. (Blum refers to the work as KD rather than J.) I can only say that in the cases that he brings I can readily imagine those formerly independent component texts, and I believe that one can see—and appreciate!—the additional layer of artistry that the editor has achieved in combining them. Indeed I have taken pains to show the ways in which the whole of the Torah is more than the sum of its parts—new and beautiful things were born out of the exquisite combination of originally exquisite pieces.[31] But one still must go a long way to prove—and not merely assert—that the artistic qualities of a text are native and that they could not possibly have been formed in a combined text.

Thus Blum, too, discounts the division of the story of Jacob's dream at Beth-El into J and E sources. He claims it has a "three-step climax":

1. "And here was a ladder standing on the earth, and its top was reaching to the heavens."
2. "And here were angels of Elohim going up and going down on it."
3. "And here was YHWH standing over him [or, for Blum: 'on it.']."

Blum asserts that this is a "perfect structure" and that "There is no reason to separate the third line for a different source and then to ascribe the perfect structure to a talented editor." No reason? There are four reasons. First, as with several other passages, this story can thus be separated into two complete, flowing stories that then can each be read as a sensible whole.[32] Second, there is a doublet of the account of Jacob's naming this place Beth-El, which is resolved into two versions, one J and one E, once we separate the two complete stories from one another.[33] Third, the nine occurrences of divine names in this story all separate into the two complete stories. And, fourth, the first three reasons *converge* with one another: the doublets resolve, the stories flow, and the names of God fit consistently into one story or the other—without any conflict or emendation of texts to make it all work. Blum must at least address such evidence. His perception that the present structure of the text has a sensible unity is not a sufficient answer to this.

Similarly, with the crucial passage concerning the introduction of the divine name at the burning bush in Exodus, Blum has argued that it is a unity and should not be divided into two originally separate texts, one J and one E. But Sean McEvenue has pointed to a series of seams in the text and strongly undermined Blum's case for unity here. And, indeed, McEvenue did so without relying on the distribution of the divine names—which would be yet again a case of convergence of these different lines of evidence. McEvenue correctly notes that "only a forced reading" could lead one to see a natural continuity here.[34]

And, once again, Blum must still come to terms with all the reverse cases, those in which the texts are obviously awkward, repetitious, or contradictory, and in which these qualities are resolved by the separation of the texts along the lines of the sources.

Methodologically, I believe that a key error of Blum's is that he has focused his analysis through the individual units of the text, and he has not sufficiently related this analysis back to the work in its final form, i.e., the work as a whole. Following the use of form criticism, especially as developed by Gunkel and von Rad, and following the use of history-of-tradition criticism *(Überlieferungsgeschichte),* especially as developed by Martin Noth and formulated by Rendtorff, Blum has concentrated on units. The Joseph story, in that view, is a "novella," the Sinai account is "self-contained." Blum separates the whole Joseph story and identifies it as an "independent composition," which is simply impossible. It is intricately bound to the rest of the Jacob story. Not only does the evidence of language reveal this, but the literary evidence as well, in that numerous events in the Jacob stories have their explicit denouements in the Joseph stories. As I discussed in the Introduction, Jacob deceives his father Isaac by using his brother Esau's cloak and the skins and meat of a goat, and years later Jacob's own sons deceive *him* using their brother Joseph's cloak and the blood of a goat. Indeed, the full chain of deceptions that begins with Jacob continues and resolves with Joseph. Blum has not come to terms with the consistent qualities that bind each of the source texts together. And so he has made crucial errors in identifying the passages that belong to each of these texts. Rendtorff, an important figure in this scholarship, formulated the method explicitly:

> . . . virtually the whole pentateuchal material is divided into such larger units: the primeval story, the patriarchal story, Moses and exodus, Sinai, sojourn in the desert, occupation of the land. Each of these units has its own characteristic profile; each is assembled from various elements of tradition and presents itself now as a more or less self-contained unit.[35]

However, the consistent characteristics of the J "units" (and of the E, P, and D units) indicate that these are not "self-contained units" at all. J was the work of an original author who composed a coherent, connected work of literature and history. This person was not an arranger or collector or synthesizer of units. The stories in J are individually interesting, intriguing, often beautiful, but they are so carefully connected to one another that one simply must come to terms with the evidence of when and how this unified work came together. Substantial evidence has been collected in the past pointing to this work's unity and antiquity, and we now have new evidence as well.

Blum, meanwhile, has attempted to account for the differences regarding the divine name by asserting that the use of sometimes the name YHWH and sometimes the generic *'ĕlōhîm* in the texts is no different from a text that sometimes says "David" and sometimes says "the King." But this does not remotely explain the *consistent* division of the thousands of occurrences of divine names in so many texts, much less the convergence of this division with the other lines of evidence. Indeed, the proof is in the pudding: one can scan the use of the terms that Blum picked as his example and readily see that in the Court History of David (see pp. 238–287) there is no pattern in the terms "the King" and "David" and "King David" that even comes close to the extraordinary consistency of the name-of-God phenomenon in the Torah.

Moreover, the issue is not simply that there are different names of God used in different parts of the Torah. Rather, the issue is over *when the name of God was revealed*. E and P develop the idea that the name YHWH is not revealed until the time of Moses. J has the name known already in the primeval history. And in texts that are attributed to E and P the name YHWH is in fact uncommon (used rarely in E and once in P) prior to the revelation of the name to Moses, but then it does become common precisely after the revelation. And, meanwhile, nothing changes in regard to the use of the name in J at that juncture. The case of the name David is not comparable to this, and Blum has not addressed the actual argument.

This matter of the divine name turns out to be a crucial point for Blum's and Van Seters's whole project. We have seen that, when J and E are separated from one another, J shows numerous signs of having been composed in the kingdom of Judah while E shows signs of composition in the kingdom of Israel. These two kingdoms existed side by side until 722 B.C.E., and so the two works J and E are traced to a period before 722. By breaking down the name-of-God distinction, Van Seters and Blum wreak havoc with the distinction between J and E—and even P. Verses that scholars thought were safely attributed to a particular work are now shuttled elsewhere. The criteria break down, the characteristic signs that associate the respective sources to Judah and Israel break down, and Blum and Van Seters are then free to assign J to a much later age.

And Van Seters and Blum never come to terms with all of the evidence that J came from Judah and E from Israel in any case. In our meeting in 1995 Van Seters observed that there still are occasionally E stories that take place in Judah rather than Israel, for example in Beersheba, and he stated that this refutes the Judah-versus-Israel provenances of J and E. But the argument was never that *every* story in J and E must take place in only one country or another. The argument was that (1) the stories in J *disproportionately* take place in Judah and the E stories in Israel; and (2) the J stories *favor* Judah while *denigrating* a site in Israel. When the spies see only Judah in J, when J pictures the acquisition of Israel's capital city of Shechem as a disgrace in Israelite history, when J describes the births of Jacob's sons only as far as the fourth son—Judah—and so on, Van Seters, Blum, and others must *address* this. Blum argued that the distinction between the accounts of those first four sons' births as belonging to J and the other sons' births as belonging to E would make the redactor of J and E into the first Bible critic! Why, he asked, would that redactor happen to use just the J stories for the first part of the narrative of the births and use just the E stories for the second part? In response to Blum's point, several possible scenarios come to mind. One possibility is that the J narrative only *had* full accounts of the first four sons, and that the author of J was as comfortable with that as he or she was with having the spies see only Judah. The redactor (who was in Judah!) chose to start with that Judean text and then added the E accounts where the J account left off. There are other possible scenarios that would explain it as well, but my point, above all, is still that the bottom line is: these distinctions *exist.* And the burden is on Blum, Van Seters, and the others to account for it. To this date, they have not attempted to do so. Robert North criticized Rendtorff, as well, for failing to come to terms with this geographical component of the Documentary Hypothesis, suggesting that "it had played no serious role in his overall analysis."[36] These scholars do not yet appreciate the force of the argument, even as they declare the model that it supports to be moribund.

Similarly, Van Seters seemed to be unaware of the fact that, when these source works in Genesis are separated from one another, they frequently each flow as a continuous story. This is one of the strongest arguments for the hypothesis. In my *Who Wrote the Bible?* I printed some of the stories in two different typesets to demonstrate this. For example, it is possible to read the J flood story and then the P flood story, and each makes sense as a complete story with no factual or grammatical breaks. When Van Seters made his new divisions of the flood stories, he destroyed the continuity of both of them with his changes. Seemingly unaware of the matter of continuity, he never justified the fragmentation that he created. In response to this criticism from me, Van Seters has tried to show that my own division of the flood story is not correct. That is a fair approach, but, first, it addresses only that one example; it is not a criticism of the entire method nor of the great mass of narrative that thus flows in stories respectively attributed to J, E, and P. Second, Van Seters's objections only add up to a small portion even of the single example of the flood story. Even when we take his case into account, there remains so much consistently flowing narrative as to call his fragmenting divisions into question. And, third, his objections are wrong in any case. For example, he observes that in my identification of the J account of the flood there is no description of the command to build the ark. He then notes that, in this J account, the deity tells Noah, "Come . . . into *the* ark" (Gen 7:1). Van Seters argues that "the command of YHWH to Noah to enter *the* ark presupposes the account of ark building in Gen 6:13–17, 22"—a text that I attribute to another source: P.[37] To Van Seters's mind, the text apparently should have said *"an* ark," not *"the* ark," if it was not referring back to this account. Van Seters has misunderstood a frequent use of the Hebrew definite article in the word *hattēbâ* here. The Hebrew definite article does not always have the same force as English "the." It corresponds to English "a" or "an" in a variety of cases. Thus, a little later in this very story the text says *wayᵉšallaḥ 'et hayyônâ,* which practically every translator has understood to mean: "And he sent out a [not *the*] dove." To cite a few other examples: השׂמלה is "a [not *the*] garment" in Gen 9:23; עין המים is "a [not *the*] pool of water" in Gen 16:7; כקיטר הכבשן means "like the smoke of a [not *the*] furnace" in Gen 19:28; הבאר is "a [not *the*] well" in Exod 2:15; and הגי is "a [not *the*] valley" in Josh 8:11.[38] Van Seters is simply mistaken. The J flood story can be read as a complete and sensible account as is. It is Van Seters who chops up the P account of the ark-building in Gen 6:13–22, thus giving to J a record of measurements in cubits (typical P!), a record that uses the word *'ĕlōhîm* in narration (typical P, but never in J). Thus he claims to have made "new divisions that restore the continuity in J"—all because of a misapprehension of the Hebrew definite article.

Van Seters does not address major webs of language and events that bind these stories together, while he out-Wellhausens Wellhausen in separating the texts on points that seem small by comparison. There is no *standard,* no clear sense of method that any reader, whether scholar or layperson, could follow. It is a series of treatments of individual stories, each on whatever particular grounds Van Seters brings in for that story. It is an ironic return to the old story-by-story approach that had been displaced by the Occam's razor of critical Bible scholarship.

Moreover, Van Seters's grounds frequently involve a kind of reasoning that leaves one in doubt. When Van Seters deals with the story of the relations between the "sons of the gods" and human women, which produce giants/Nephilim (Gen 6:1–4), he compares the story to the Greek *Catalogue of Women* of Hesiod. Because the *Catalogue of Women* has parallels to the biblical story but does not include the element of giants, Van Seters concludes that "it would be best to view the reference to the Nephilim in Gen 6:4 as secondary."[39] Again with due respect, this is weak reasoning. The presence of an element in the Bible's story that is not present in the Hesiodic work does not make that element "secondary," yet Van Seters arrives at this conclusion without further defense or comment.[40]

The way Van Seters comes to date the story of the tower of Babylon is another case of doubtful reasoning, thus: He says that the Bible's story of the tower must be based on the ziggurat Etemenanki in Babylon. He says that the earliest reference to Etemenanki is in the *Erra* epic and that *Erra* is dated ca. 765 B.C.E. by von Soden. He says that there are numerous references to the Babylonian temple Esagila, but not to the ziggurat Etemenanki before this. He concludes: "this silence about Etemenanki can only mean that it probably did not predate the eighth century B.C.E."[41] He then adds that it is not likely that the ziggurat was an "exceptional structure" until much later. Therefore the tower of Babylon story must have been written much later—close to the time of the exile. This is troubling in many ways. Van Seters does not acknowledge that the date of the *Erra* epic is a contentious issue. He does not make clear that the first reference to the ziggurat Etemenanki *by name* is in the *Erra* epic but that the ziggurat of Babylon is in fact mentioned earlier than that. It is mentioned in the *Enuma Elish.* And it is certainly understood to be an "exceptional structure" there (Tablet VI, line 63). And Van Seters's argument is a weak argument from silence in any case.

There are other serious problems of method in Van Seters's case that J was composed in the Babylonian exile. He claims that the themes of exile and dispersion early in Genesis (for example, in the tower of Babylon story) "point very clearly to the concerns of the exilic community." He says this despite repeated warnings by biblical scholars against thinking that references to exile in the Bible had to be written by someone who had actually experienced exile. Exile of defeated nations was a reality in the ancient Near East, and an author did not have to be carried off in chains in order to think of writing a story of humankind being dispersed from the tower of Babylon.

Another doubtful reasoning is involved in Van Seters's treatment of the entire matter of the promises to the patriarchs. Van Seters makes a number of connections between these accounts of Abraham, Isaac, and Jacob, on one hand, and accounts of David and Solomon, on the other. He concludes that the author of the patriarchal stories has taken the language and ideology of monarchy and transplanted them into the patriarchal stories. One might take this as evidence that the patriarchal stories were written in the period of the monarchy, but Van Seters concludes the opposite: that the patriarchal stories must have been written after the monarchy was ended. Why? "Because there is no effort in the biblical tradition to make a linear connection between the patriarchs and David until very late."[42] First of all, that is no argument. Second, it is just not true in any case. Jacob's deathbed blessing, a very early biblical text, which I mentioned earlier, makes such a connection between patriarchs and monarchy (Gen 49:8–10). The end of the Judah-and-Tamar story (Genesis 38) also makes the connection. Van Seters's argument is classically circular. He is trying to argue for the lateness of J, and, at the same time, he takes the lateness of J to be proof of his argument. He says on that same page, "The lateness of the texts also confirms this interpretation." But the lateness of the texts was what he was trying to *prove* with this interpretation.[43]

There is, therefore, reason to doubt the soundness of method and reasoning in Van Seters's work. In this scholarship the text rarely speaks for itself. Rarely is the point automatically manifest in the text that Van Seters puts before us for our consideration. Rather it is the scholar's spin on the text that houses the point. This is all the more disturbing because Van Seters frequently does not come to terms with whole areas of evidence and scholarship that are relevant to his analyses of the text. He traces a development in the conditionality of the possession of the land, and he discusses the relationship between the biblical covenants, as if he were unaware of the substantial work in the field on the controversial question of conditionality of the biblical covenants. If the Abrahamic covenant is an unconditional promise of land to Abraham's descendants, and the Sinai covenant makes possession of the land conditional upon obedience to the commandments, how are these two covenants to be reconciled? There is a mountain of interesting work on this by a number of scholars, which Van Seters does not so much as mention.

The same goes for the arguments on various other grounds for the early date of the work known as P. Van Seters acknowledges that P was written later than J, and so he must come to terms with a fairly large group of scholars who have brought evidence and arguments that even P was pre-exilic. Besides my own work and the works by the linguists whom I have mentioned, there is David Noel Freedman's important work on the Primary History, arguing that all of the sources of the Torah were completed by the exile.[44] Menahem Haran, as well, following on the massive work of his teacher, Yehezkel Kaufmann, dates P to around the same time that I do. Van Seters refers to Haran's articles on book scrolls and literacy but does not come to terms with Haran's considerable work on the earliness of P.[45] One cannot ignore whole corpora of mainstream scholarship that are counter to one's own. If P was pre-exilic and J was earlier than P, then Van Seters's late dating of J is impossible.

Van Seters simply does not come to terms with the present state of scholarship in some areas that are relevant and important to his research.

I have already referred to his not dealing with the matter of the names of God in terms of the present state of the evidence and to his not dealing with the matter of the continuity of narratives. Van Seters proposes a picture of P as "dependent upon and an addition to J,"[46] but if P had been written as a set of supplemental lines around J stories, how is it possible that P flows as a continuous narrative when it is separated from J? That is, the considerable continuity of the P stories indicates that they were originally a separate work, independent, not an addition to anything. His criticism of my own division of the flood story does not begin to come to terms with this problem. I also referred to his not dealing with the full evidence that J came from Judah and E from Israel, which seriously undermines his proposal. For example, when he looks at the story of the births and naming of Jacob's twelve sons, he says that "Efforts to divide it into . . . J and E are hardly convincing and have been rightly rejected,"[47] and he concludes that this story has nothing to do with any political reckoning of the twelve tribes. As we have seen, however, the section that uses the name YHWH covers three tribes that had no territory (Reuben, Simeon, and Levi—the same three who lose out in the J stories I cited above) and then culminates in *Judah*. And the section that refers to God as *'ĕlōhîm* covers only the tribes of the kingdom of *Israel*. Van Seters cannot simply ignore such an obvious piece of evidence against his proposal and write it off as "hardly convincing."[48]

Likewise in the matter of his dating J later than the Greek historians, Van Seters has not come to terms with the contrary evidence. His response to Baruch Halpern's work is of particular interest since it so challenges his own. Van Seters writes in a note that "The difference between ancient and modern historians is not understood by him [Halpern]. He simply ignores the whole subject of ancient historiography outside of the Old Testament."[49] And Van Seters says in another note at the beginning of this book, "Halpern's book offers a major critique to my own work in ancient historiography, so the present work may be viewed as an implicit response to it."[50] In the first place, the charge that Halpern ignores ancient historiography outside of the Bible is incomprehensible. Halpern has referred to the ancient Greek historians more than almost anyone in the field; he contributed to a conference and resulting book that I edited involving this subject some years ago that brought together biblical, modern, and classical historians, including the preeminent classical historian Arnaldo Momigliano;[51] and he has regularly acknowledged his intellectual partnership in his work with his father, who was a classical historian. More to the present point, though, Halpern's work is too significant and too rigorously argued to be written off so casually. Halpern is one of the most sophisticated scholars of historiography in the history of our field, and to say that he does not understand the difference between ancient and modern historians is an unfortunate charge where a serious, responsible defense was called for—and not just a promise of an "implicit" answer.

This is all the more serious since even Blum has questioned Van Seters's picture of Israelite history writing emulating that of the Greeks. In our meeting, Blum made the point that Herodotos and Hekataios had a wholly different notion of history, as evidenced from the very first words of their works. I made the same point on the first page of the conference volume that I mentioned above, namely, that Herodotos and Hekataios begin their works by introducing themselves and stating their purposes:

> This book is the result of the historical inquiries *(historiê)* of Herodotos of Halikarnassos. Its purpose is twofold . . .

> Hekataios of Miletos writes this. I write the following things believing them to be true; for the things said among the Greeks seem to me numerous and absurd.

It is hard to imagine anyone reading these and then reading J and feeling that he or she is in familiar territory.[52]

In any case, if Van Seters has any chance of making the case for his proposal, he must come to terms with the mass of evidence that contradicts it, and he must address the mass of scholars who disagree with him. It is such a strange proposal that it requires the most careful argumentation and attention to the evidence and arguments of others. What evidence compels him to see everything so late? To think that the Jews produced so little of this writing until they were evicted from their land? To see the centuries in which they flourished in their land as being so barren of literary-historical production? To have such a negative view that "the unifying factor in the national history of Israel" was not their covenants or their beliefs or their monarchy, but "the people's sin and the divine judgment"? To see these late writers as inventing so much of their history? These matters require a substantial, meticulous defense, but overall his work has a much more limited scope: putting a new proposal on the table. That proposal is insufficiently defended thus far and, in my judgment, ultimately indefensible.

Some people describe the present state of studies of biblical authorship as being in disarray. They see the former consensus about the main elements of the Documentary Hypothesis breaking down. I have to say that the rash of new proposals and models does not reflect a weakness in the hypothesis; it rather owes to a breakdown of standards of method, starting with a failure to come to terms with the case for the view that they think no longer prevails. The evidence is not in disarray. What is in disarray is methodology. Those who have challenged the antiquity of J have not adequately substantiated their assertion. And, in any case, they have not taken on the combination of evidence that is the basis for the theory that has dominated our field for over a hundred years. And now the identification of the work of the first great writer provides new evidence that adds further confirmation of the distinction between two sources, J and E, and further confirmation that these source works must derive from the period of the divided monarchy. They are the oldest known works of prose.

4. DISTRIBUTION OF TERMS IN PROSE NARRATIVE

	J (Gen-Deut)	Joshua	Judges	1 Samuel (B)	2 Samuel	All Other Narrative
foolish נבל, נבלה	Gen 34:7	7:15	19:23, 24; 20:6, 10	25:25	3:33; 13:12, 13	none
Why did you deceive me? למה רמיתני	Gen 29:25	9:22		19:17	19:27	none
with deception במרמה	Gen 27:35; 34:13			28:12		none
fodder מספוא	Gen 24:25, 32; 42:27; 43:24		19:19			none
slumber descended תרדמה נפלה	Gen 2:21; 15:12			26:12		none
content to live ויואל לשבת	Exod 2:21	7:7	17:11; 19:6			none
his eye was dim כהתה עינו	Gen 27:1; Deut 34:7			3:2		none

	J (Gen-Deut)	Joshua	Judges	1 Samuel (B)	2 Samuel	All Other Narrative
regularly מימים ימימה			11:40; 21:19	1:3; 2:19		none[1]
kindness and faithfulness חסד ואמת	Gen 24:27, 49; 47:29; Exod 34:6	2:14			2:6; 15:20	none
wash feet רחץ רגלים	Gen 18:4; 19:2; 24:32; 43:24		19:21	25:41	11:8	none
resident of the land יושב הארץ	Gen 50:11; Exod 34:12, 15; Num 14:14		11:21		5:6	none
Sheol שׁאול	Gen 37:35; 42:38; 44:29, 31; Num 16:30, 33			2:6	1 Kgs 2:6, 9	none
shear גזז	Gen (31:19a[2]); 38:12, 13			25:2, 4, 7, 11	13:23, 24	none
turn right or left נטה ימין ושׂמאל	Num 20:17; 22:26[3]				2:19, 21	none

	J (Gen-Deut)	Joshua	Judges	1 Samuel (B)	2 Samuel	All Other Narrative
birds of prey עיט	Gen 15:11			15:19; 25:14		none
old, well along in days זקן בא בימים	Gen 18:11; 24:1	13:1 (23:1)[4]			1 Kgs 1:1	none
conceal צפן	Exod 2:2f.	2:4				none
And there was a man from ויהי איש אחד מן			13:2	1:1		none
coat of many colors כתנת פסים	Gen 37:3				13:18	none
Take off your shoes . . . . . . של נעלך	Exod 3:5	5:15				none
spies מרגלים	Gen 42:9, 11, 14, 16, 30, 31, 34	2:1; 6:22, 23		26:4	15:10	none

	J (Gen-Deut)	Joshua	Judges	1 Samuel (B)	2 Samuel	All Other Narrative
in front of the sun נגד השמש	Num 25:4				12:12	none
hair grows out *Piel of* צמח			16:22		10:5	(1 Chr 19:5)[5]
reject נאץ	Num 14:11, 23 16:30			2:17	12:14	Deut 31:20
to spy *(verbal forms)* רגל	Num 21:32	6:25; 7:2; (14:7)	18:2, 14, 17		10:3; 19:28	(1 Chr 19:3); Deut 1:24[6]
opened mouth פצתה פה	Gen 4:11; Num 16:30		11:35, 36			Deut 11:6[7]
roof (as setting) גג		2:6(bis), 8	9:51; 16:27		11:2; 16:22; 18:24	1 Sam 9:25f. (A)

	J (Gen-Deut)	Joshua	Judges	1 Samuel (B)	2 Samuel	All Other Narrative
know (with sexual connotation) ידע	Gen 4:1, 17; 19:8; 24:16; 38:26		11:39; 19:25; 21:11, 12	1:19	1 Kgs 1:4	Gen 4:25 (R)
know (with double connotation[8]*)* ידע	Gen 2:9, 17; 3:7; 19:5, 33, 35; 38:16; 39:6, 8				17:8; 1 Kgs 1:11, 18[9]	Ruth 3:3, 4
lie with (with sexual connotation) שכב	Gen 19:32, 33, 34, 35; 26:10; 34:2, 7; 35:22; 39:7, 10, 12	2:1, 8[10]	16:3	2:22	11:4, 9, 11, 13; 12:3, 11, 16, 24; 13:5, 6, 8, 11, 14, 31; 1 Kgs 1:2	Gen 30:15f. (E)
virgin בתולה	Gen 24:16		11:37, 38; 19:24;		13:2, 18; 1 Kgs 1:2	Esth 2:2, 3, 17, 19

	J (Gen-Deut)	Joshua	Judges	1 Samuel (B)	2 Samuel	All Other Narrative
Don't hide it from me אל תכחד ממני		7:19		3:17	14:18	Jer 38:14, 25
attractive figure יפת תאר	Gen 29:17; 39:6			25:3	1 Kgs 1:6[11]	Gen 41:18 (E)
beautiful, handsome יפת מראה	Gen 29:17; 39:6				11:2; 14:27	Gen 41:2 (E); 1 Sam 17:42 (A)
perform שחק			16:25, 27	18:7	2:14; 6:5, 21	(1 Chr 13:8); (15:29); 30:10
crushed, struggled רצץ	Gen 25:22		9:53; 10:8	12:3, 4		2 Kgs 18:21 (= Isa 36:6); 2 Chr 16:10
cursed ארור	Gen 3:14, 17; 4:11; 5:29; 9:25; 12:3; 27:29 (bis); Num 22:6, 12[12]		21:18	26:19		1 Sam 14:24, 28 (A); 2 Kgs 9:34

	J (Gen-Deut)	Joshua	Judges	1 Samuel (B)	2 Samuel	All Other Narrative
drunk שׁכר	Gen 9:21; 43:34			1:13, 14; 25:36	11:13	1 Kgs 16:9; 20:16
old age, gray hair שׂיבה	Gen 25:8; 42:38; 44:29, 31		8:32		1 Kgs 2:6, 9	Gen 15:15 (R); Ruth 4:15
teraphim תרפים			17:5; 18:14, 17, 18, 20	15:23; 19:13, 16		Gen 31:19, 34f. (E); 2 Kgs 23:24
good-for-nothing בליעל			19:22; 20:13	1:16; 2:12; 10:27; 25:17, 25; 30:22	16:7; 20:1	1 Kgs 21:10, 13; 2 Chr 13:7
flesh and bone בשׂר ועצם	Gen 2:23 29:14		9:2		5:1 19:13, 14	[1Chr 11:1]; Job 2:5
kept quiet החריש	Gen 24:21; 34:5; Exod 14:14		16:2; 18:19	7:8; 10:27	13:20; 19:11	2 Kgs 18:36; Esth 4:14; 7:4

	J (Gen-Deut)	Joshua	Judges	1 Samuel (B)	2 Samuel	All Other Narrative
(news) בשׂר				4:17; 31:9	4:10; 18:19, 20 22, 25, 26, 27, 31; 1 Kgs 1:42	2 Kgs 7:9; [1 Chr 10:9]
anguished עכר	Gen 34:30	6:18; 7:24, 25, 26	11:35			1 Sam 14:29 (A); 1 Kgs 18:17f.; (1 Chr 2:7)[13]
hate שׂנא	Gen 24:60; 26:27; 29:31, 33; 37:5, 8; Num 10:35		14:16; 15:2		5:8; 13:15, 22; 19:7	Gen 37:4; Deut 1:27; 9:28;[14] 1 Kgs 22:8; (= 2 Chr 18:7)
suffering עצב	Gen 3:16 (bis), 17; 5:29; 6:6; 34:7; 45:5				19:3; 1 Kgs 1:6	1 Sam 20:3, 34; 1 Chr 4:9f.; Neh 8:10f.

	J (Gen-Deut)	Joshua	Judges	1 Samuel (B)	2 Samuel	All Other Narrative
and he sent messengers וישלח מלאכים	Gen 32:4 Num 20:14; 21:21; 22:5[17]	7:22	9:31; 11:12, 14, 17, 19[16]	6:21; 11:3; 16:19; 19:11, 14, 20, 21	2:5; 3:12, 14, 26; 5:11; 11:4; 12:27	Deut 2:26;[15] 1 Kgs 20:2; 2 Kgs 1:2, 16; 14:8; 16:7; 17:4; 19:9; (1 Chr 14:1); 19:2; 2 Chr 35:21
expand, spread פרץ	Gen 28:14; 30:30, 43; 38:29; Exod 19:22, 24			3:1; 25:10; 28:23	5:20; 6:8; 13:25, 27	Exod 1:12 (E); 2 Kgs 5:23; 14:13; 17x in Chr
went on again יסף עוד	Gen 18:29; 37:5, 8; 38:26		9:37; 11:14; 13:21	3:6; 27:4	2:22; 7:20; 18:22	Num 22:15;[18] 1 Sam 23:4

NOTES ON DISTRIBUTION OF TERMS

1. The phrase occurs in Exod 13:10, but the passage is law, not prose narrative.
2. This verse falls at the juncture where three sources meet and is therefore not certain. (E = Gen 31:1–2, 4–16, 19bff.; J = 3, 17; P = 18.)
3. Num 22:26 is part of the Balaam story. I have not included it in the work because I have never felt I could confidently separate the J and E components of the story. See my note, p. 315. Among those who do separate J and E in the Balaam story, though, the consensus is that this verse is part of J. See Jo Ann Hackett, "Balaam," *Anchor Bible Dictionary* 1: 569.
4. The phrase in Josh 23:1 is an epanalepsis (also known as resumptive repetition or *Wiederaufnahme*) of the passage in 13:1. As an editorial device of the Deuteronomistic historian it is not correctly regarded either as another occurrence of the phrase in this work or as an occurrence belonging to another work.
5. Verses in Chronicles that repeat verses in Samuel or Kings are bracketed. The Samuel and Kings verses are presumably the Chronicler's source.
6. Deut 1:24 appears to be based on Num 13:23 (J).
7. Deut 11:6 appears to be quoting Num 16:30 (J).
8. This refers to verses in which the verb "to know" *(yd')* has a double sense or paronomasia, playing on the sexual and nonsexual meanings of the word.
9. These references to the king's not knowing here in 1:11, 18 appear to be a pun on the king's not knowing Abishag in 1:4.
10. The verb in Joshua may mean simply to lie down, but since it takes place in a brothel I take it to be a double entendre.
11. 1 Kgs 1:6 has טוב תאר rather than יפת תאר
12. On the verses that are part of the Balaam story in Numbers 22–24, see note 3.
13. The verse in Chronicles refers to the violation of *ḥerem* in Joshua 6–7, which is a J narrative.
14. Other uses of the word in Exodus, Deuteronomy, and Joshua occur in legal formulae, particularly with respect to cities of refuge, referring to those who took life without prior hate of the victim.

15. Deut 2:26 appears to be quoting Num 21:21 (J).
16. Contrast Judg 6:35; 7:24, which come from a different source and use a different formation of the phrase.
17. See note 3. This verse, too, is seen as part of J by those who divide the Balaam story.
18. See note 3.

NOTES

INTRODUCTION

1. R. E. Friedman, *The Exile and Biblical Narrative,* Harvard Semitic Monographs (Atlanta: Scholars Press, 1981), pp. 44–119; "Torah," *Anchor Bible Dictionary* 6: 609–19; *Who Wrote the Bible?,* 2nd ed. (San Francisco: Harper San Francisco, 1997), pp. 161–216 (first edition: New York: Simon & Schuster, Summit, 1987).
2. Peter Ellis, *The Yahwist: The Bible's First Theologian* (Notre Dame, IN: Fides, 1968); Robert Coote and David Ord, *The Bible's First History* (Philadelphia: Fortress, 1989); Harold Bloom and David Rosenberg, *The Book of J* (New York: Grove Weidenfeld, 1990).
3. Erich Auerbach, *Mimesis* (Princeton, NJ: Princeton University Press, 1968), p. 12; German edition, 1946.
4. Robert Polzin, *Late Biblical Hebrew: Toward an Historical Typology of Biblical Hebrew Prose,* Harvard Semitic Monographs (Atlanta: Scholars Press, 1976). Polzin's analysis showed that E came from this period as well.
5. See the Appendix note 24 (p. 399) for citations.
6. The similarity between Judah's wife, who is called the daughter of Shua, and David's wife is so close that the two are reversed in Chronicles, apparently the only clear waw/vet spelling reverse in the Hebrew Bible. Many other parallels of names between these stories have been noted. See J. A. Emerton, "An Examination of a Recent Structuralist Interpretation of Genesis XXXVIII," *Vetus Testamentum* 26 (1976): 96; "Some Problems in Genesis XXXVIII," *Vetus Testamentum* 25 (1975): 344; Gary Rendsburg, "David and His Circle in Genesis XXXVIII," *Vetus Testamentum* 36 (1986): 438–46.
7. My colleague David Noel Freedman pointed out to me that this is the case in the light of text-critical analysis.
8. Lawrence Boadt, *Reading the Old Testament: An Introduction* (New York: Paulist Press, 1984), p. 98: "J has to be traced to the great outburst of energy and talent that flowered in the new empire created by David and Solomon. Several other parts of the Bible come from this same period and share many of the artistic merits of the Yahwist. There is, for example, the wonderful story of the conflicts among David's sons for the right of succession to the throne that we find in 1 Samuel 9–20 and 1 Kings 1–2. It is a masterpiece of psychological insight into David the mighty ruler who was too soft on his own sons. A very similar talent for portraying the weakness and strengths of Saul can be found in the story of David's own rise to power in 1 Samuel 16–31. These both have been written in the style seen in J. . . ."

9. Joel Rosenberg, *King and Kin* (Bloomington, IN: Indiana University Press, 1986), p. xiii.
10. Yair Zakovitch, "Assimilation in Biblical Narratives," in J. Tigay, ed., *Empirical Models for Biblical Criticism* (Philadelphia: University of Pennsylvania Press, 1985), pp. 185–92.
11. Ronald Clements, *Abraham and David* (London: SCM, 1967).
12. Bloom, *The Book of J.*
13. The literary factors include the many doublets, the contradictions in the text that resolve when Samuel A and B are separated, and the quantity of narrative continuity of each source that is revealed when the two are separated.
14. Except in 1 Samuel 17–18, where the Greek text indicates serious textual complications.
15. The Greek text and all English translations that follow its order of the biblical books include the book of Ruth after the book of Judges. Ruth is located elsewhere in the Hebrew (Masoretic) text and was not part of the continuous narrative of the Deuteronomistic history. It does not figure in the present analysis.
16. Baruch Halpern, *The Constitution of the Monarchy in Israel,* Harvard Semitic Monographs (Atlanta: Scholars Press, 1981), chap. 4. More specifically, the division comes at the last couple of verses of Judges 8.
17. Hebrew *ḥesed we'ĕmet.*
18. Hebrew *škb.*
19. Hebrew *wytmhmh.*
20. The iconography of the Pentateuchal sources reflects the time and locale in which each was produced. See the Appendix. See also *Who Wrote the Bible?,* pp. 73–76.
21. The words "I swear" are used in the English here to reflect the Hebrew terms that constitute an oath formula: "as I live" in v. 21, and Hebrew *'im* in v. 23.
22. A third wife/sister story, which occurs in E, makes a point of saying that the king does not touch the woman (Sarah, Genesis 20).
23. The story of the war between Benjamin and the rest of the Israelite tribes is also presented in terms of brothers-killing-brothers (Judg 20:13, 23, 28; 21:6); and there, too, the word "field" comes up twice.

    J. Blenkinsopp noted the parallel references to "field" in Gen 4:8 and 2 Sam 14:6 in "Theme and Motif in the Succession History (2 Sam xi 2ff.) and the Yahwist Corpus," *Volume du Congrès: Genève 1965,* Supplements to *Vetus Testamentum* 15 (1966): 51. He also pointed out a number of additional parallels of theme between J and the Court History.

24. E. A. Speiser, *Genesis, Anchor Bible* (Garden City, NY: Doubleday, 1964), p. 227; Nahum Sarna, *Understanding Genesis* (New York: Schocken, 1966), p. 184; Michael Fishbane, *Text and Texture* (New York: Schocken, 1979), pp. 55ff.; Robert Alter, "Sacred History and Prose Fiction," in R. E. Friedman, ed., *The Creation of Sacred Literature* (Berkeley, CA: University of California Press, 1981), p. 23. Fishbane's discussion of recompense for deception is especially interesting in parallel with what follows.
25. I first identified these cases in an article, "Deception for Deception," *Bible Review* 2, no. 1 (1986): 22–31, 68, without relating them to the matter of the hidden book in the Bible.
26. David Noel Freedman, "Dinah and Shechem, Tamar and Amnon," in *God's Steadfast Love: Essays in Honor of Prescott Harrison Williams, Jr.*, Austin Seminary Bulletin 105:2 (1990), pp. 51–63; also published in Freedman, *Divine Commitment and Human Obligation* (Grand Rapids, MI: Eerdmans, 1997), pp. 485–95.
27. In 2 Sam 3:25, Joab tells David that Abner came *l*^{e}*pattōt*^{e}*kā* (the same term that is used elsewhere in the work).
28. I do not mean to enter here upon the entire debate over when Israel first became monotheistic. I do think that monotheism was already present in Israel at the time this work was composed (ninth century B.C.E.). I cannot say what proportion of the people followed it at that time. In any case, it does not change my present point, that the author depicted the deity as male.

AFTERWORD

1. See the Introduction, pp. 35–36.
2. The closest is when "YHWH's word came to Nathan" in 2 Samuel 7, but even this is indirect in that the message is for David and is here conveyed by words coming to a prophetic intermediary.
3. Richard Elliott Friedman, *The Hidden Face of God* (San Francisco: Harper San Francisco, 1997), pp. 7–59; originally published as *The Disappearance of God* (New York: Little, Brown, 1995).
4. Peter Ellis, *The Yahwist: The Bible's First Theologian* (Notre Dame, IN: Fides, 1968); Robert Coote and David Ord, *The Bible's First History* (Philadelphia: Fortress, 1989); Harold Bloom and David Rosenberg, *The Book of J* (New York: Grove Weidenfeld, 1990).

APPENDIX

1. The phrase (ויואל לשבת) occurs in Josh 17:12 and Judg 1:27, 35, which are not part of this work, but it has a manifestly different meaning in these passages. The striking difference between the way the phrase is understood in this work and the way it is used in these other places is reflected in the fact that virtually all Bible translations convey it differently in the two groups.
2. Num 20:17 and 2 Sam 2:19, 21 are certain. Num 22:26 is part of the Balaam story, which has always been regarded as one of the most complex portions of the Hebrew Bible for source analysis. In my *Who Wrote the Bible?* I grouped it with E even though I recognized that a more complex literary history was present, and I have not included it in the present collection because I still have not solved the problem of its literary history. But scholars who do divide the Balaam story into J and E have almost without exception given Num 22:26 to J. So S. R. Driver, *Introduction to the Literature of the Old Testament* (orig. ed., 1891; Gloucester: Peter Smith, 1972), and Martin Noth, *A History of Pentateuchal Traditions* (Ger. ed., 1948; Englewood Cliffs, NJ: Prentice-Hall, 1972); cf. Hackett, "Balaam," *Anchor Bible Dictionary* 1: 569.

   Note that the Deuteronomist regularly uses a different verb, the root *swr,* rather than *nṭh,* in referring to turning left or right (Deut 2:27; 5:29; 17:11, 20; 28:14; Josh 1:7; 23:6; 2 Kgs 22:2).
3. It occurs in a passage of law in Exod 13:10.
4. *Who Wrote the Bible?* 2d ed. (San Francisco: Harper San Francisco, 1997), pp. 73–76; "Torah (Pentateuch)," *Anchor Bible Dictionary* 6: 612 (item 11), 613 (item 11).
5. It is mentioned in Samuel B, but precisely in identifying it with Shiloh (1 Sam 2:22), never before or after.
6. Baruch Halpern, *The First Historians* (San Francisco: Harper & Row, 1988).
7. David Noel Freedman, "Dinah and Shechem, Tamar and Amnon," in *God's Steadfast Love: Essays in Honor of Prescott Harrison Williams, Jr.,* Austin Seminary Bulletin 105:2 (1990), pp. 51–63; also published in Freedman, *Divine Commitment and Human Obligation* (Grand Rapids, MI: Eerdmans, 1997), pp. 485–95.
8. In the E story of Moses' Cushite wife, the wife does not act. And in E Moses is not pictured as having siblings. (Only in P are Miriam and Aaron identified as his sister and brother.)
9. See J. A. Emerton, "Some Problems in Genesis XXXVIII," *Vetus Testamentum* 25 (1975): 344. Emerton treats the many connections between Genesis 38 and the tribe of Judah, pp. 343–46; cf. Emerton, "Judah and Tamar," *Vetus Testamentum* 29 (1979): 404ff. He also

demonstrates the weakness of attempts that some scholars had made to argue that Genesis 38 was separate from J or an addition to it.

10. G. Rendsburg, "David and His Circle in Genesis XXXVIII," *Vetus Testamentum* 36 (1986): 438–46. Rendsburg deserves credit for pointing out several parallels of names between J and the Court History. R. E. Clements, *Abraham and David* (London: SCM, 1967); W. Brueggemann, "David and His Theologian," *Catholic Biblical Quarterly* 30 (1968): 156–81. Cf. Benno Jacob, *Genesis: Das erste Buch der Tora* (Berlin, 1934), pp. 1048–49.
11. J. Blenkinsopp analyzed parallels of theme in the Court History and J and wrote that "Stylistic analysis may *suggest* the work of one author beneath the present redaction of the Yahwist strand and the Succession History but it cannot hope to *establish* more than a homogeneous development in literary type and expression." He was correct on both counts. It does *suggest* single authorship, but it cannot prove it alone. Blenkinsopp concluded: "this study of theme and motif must be pursued together with detailed stylistic analysis—vocabulary, word-order, deployment of dialogue, grammatical and syntactic peculiarities, paronomasia, constructional techniques . . ." Now that most of these categories have been examined here and in the linguistic studies I have cited, the result is that they do in fact converge in a way that confirms the suggestion of common authorship. Blenkinsopp, "Theme and Motif in the Succession History (2 Sam xi 2ff.) and the Yahwist Corpus," *Volume du Congrès: Genève 1965,* Supplements to *Vetus Testamentum* 15 (1966): 50n., 57.
12. E. C. Steuernagel, *Die Bücher Samuel* (1926). Cf. Blenkinsopp, "Theme and Motif in the Succession History," pp. 44–57.
13. Budde viewed J as extending to 1 Kings 2 but did not attribute it to a single author. D. Karl Budde. *Die Biblische Urgeschichte* (Giessen: J. Ricker, 1883); *Die Bücher Richter und Samuel: Ihre Quellen und ihr Aufbau* (Giessen: J. Ricker, 1890). The evidence of characteristic terms and phrases runs contrary to Budde's source identifications.
14. See Blenkinsopp, "Theme and Motif in the Succession History," pp. 44–45, for discussion and references.
15. M. Noth, *The Deuteronomistic History,* trans. J. Doull (Sheffield: JSOT, 1981), p. 55; original German edition, *Überlieferungsgeschichtliche Studien* (Tübingen: Max Niemeyer, 1943). Cf. P. K. McCarter, *II Samuel, Anchor Bible* (New York: Doubleday, 1984), pp. 6f.
16. R. E. Friedman, *The Exile and Biblical Narrative*, Harvard Semitic Monographs (Atlanta: Scholars Press, 1981), pp. 1–43; "From Egypt to Egypt: Dtr1 and Dtr2," in *Traditions in Transformation: Turning-Points in Biblical Faith*, Frank Moore Cross *Festschrift,* eds. B. Halpern and J. Levenson (Winona Lake, IN: Eisenbrauns, 1981), pp. 167–81; *Who*

Wrote the Bible? 2d ed. (San Francisco: Harper San Francisco, 1997), pp. 101–49; "Torah," *Anchor Bible Dictionary* 6 (1992): 605–22; "The Deuteronomistic School," in *Fortunate the Eyes That See,* David Noel Freedman *Festschrift,* ed. A. Beck, et al. (Grand Rapids, MI: Eerdmans, 1995), pp. 70–80.

17. Those who claim that the Deuteronomistic historian "invented" the history following the exile simply ignore the linguistic evidence that shows the Court History (and J as well) to have been written in Classical Biblical Hebrew. This evidence utterly undermines their position. See below.

18. Baruch Halpern, *The Constitution of the Monarchy in Israel,* Harvard Semitic Monographs (Atlanta: Scholars Press, 1981), chap. 4.

19. *The Exile and Biblical Narrative,* pp. 44–119; *Who Wrote the Bible?* pp. 161–216.

20. **E:** Gen 20:1–18; 21:6, 8–34; 22:1–10, 16b–19; 25:1–4; 28:11b, 12, 17, 18, 20–22; 30:1b–3, 4b–24a; 31:1–2, 4–16, 19b–54; 32:1–3, 14b–33; 33:1–18aα, 18b–20; 35:1–8, 16aβ–20; 37:3a, 4, 12–18, 21–22, 24, 25a, 28a, 29, 30, 36; 40:1–23; 41:1–45a, 46b–57; 42:5–7, 21–25, 35–37; 43:14, 23b; 45:3; 46:1–5a; 47:7–10; 48:1–2, 8–22; 50:15–22a, 23–26; Exod 1:8–12, 15–21; 3:1, 4b, 6, 9–18; 4:1–18, 20b, 21a, 22, 27–31; 5:3–23; 6:1; 7:14–18, 20b–21a, 23–29; 8:3b–11a, 16–28; 9:1–7, 13–34; 10:1–19, 21–29; 11:1–8; 12:21–27, 29–36, 37b–39; 13:1–19; 14:5b, 7, 11–12, 19a, 20a, 25a; 15:20–21, 25b–26; 17:2–16; 18:1–27; 19:2b–9, 16b–17, 19; 20:18–26; 24:1–15a, 18b; 32:1–35; 33:1–23; Num 11:1–35; 12:1–16; 21:4b–9; Deut 31:14–15, 23

    J: Gen 2:4b–25; 3:1–24; 4:1–24, 26b; 6:1–8; 7:1–5, 7, 10, 12, 16b–20, 22–23; 8:2b–3a, 6, 8–12, 13b, 20–22; 9:18–27; 10:8–19, 21, 23–30; 11:1–9; 12:1–4a, 6–20; 13:1–5, 7–11a, 12b–18; 15:6–12, 17–21; 16:1–2, 4–14; 18:1–33; 19:1–38; 21:1a, 2a, 7; 22:20–24; 24:1–67; 25:8a, 11b, 21–34; 26:1–33; 27:1–45; 28:10, 11a, 13–16, 19; 29:1–35; 30:1a, 4a, 24b–43; 31:3, 17, (19a); 32:4–14a; 33:1a, 3b, 4, 16; 34:1–31; 35:21–22; 36:31–43; 37:2b, 5–11, 19–20, 23, 25b–27, 28b, 31–35; 38:1–30; 39:1–23; 42:1–4, 6, 8–20, 26–34, 38; 43:1–13, 15–23a, 24–34; 44:1–34; 45:1–2, 4–28; 46:5b, 28–34; 47:1–6, 11–27a, 29–31; (49:1–27); 50:1–11, 14; Exod 1:6, 22; 2:1–23a; 3:2–4a, 5–6, 19–22; 4:19–20; 24–26; 5:1–2; 13:21–22; 14:5a, 6, 10bα, 13–14, 19b, 20b, 22aβ, 24, 25b, 27b, 30–31; 15:22b–25a; 16:4–5, 35b; 19:10–16a, 18, 20–25; 34:1a, 2–28; Num 10:29–36; 13:17–20, 23–24, 27–31, 33; 14:1b, 4, 11–25, 39–45; 16:1b, 2a, 12–14, 25, 27b–32a, 33–34; 20:14–21; 21:1–3, 21–35; 25:1–5; Deut 34:1–7.

21. Also, when Abraham is told to migrate from Mesopotamia he is told in Gen 12:3 that all the families of the earth will be blessed through him

(J); but then in Gen 22:18 he is told anew that all the nations will be blessed through him, this time as a reward for his listening to God in the matter of the sacrifice of Isaac (E).

22. Gen 21:1. The other, well-known possible exception is the occurrence of the divine name in Gen 17:1, and I do not think that even this single case is an exception. In Exod 6:3 the deity tells Moses that His name is YHWH but that He made Himself known to the patriarchs only as El Shadday. Gen 17:1 makes a special point of conveying this by saying: "And *YHWH* appeared to Abraham and said to him, 'I am *El Shadday.*'" This strikes me as the rule, not the exception.

23. I understand Reuben to have ceased to exist as an independent tribe by the time of the composition of J and of E. It lay in the most vulnerable geographical position of all the tribes: bordering Moab and Ammon but cut off by the Dead Sea from physical contact with the tribes that were west of the Jordan River. It is mentioned among the tribes in the early Song of Deborah, but it ceased to exist on its territory early on.

24. Robert Polzin, *Late Biblical Hebrew: Toward an Historical Typology of Biblical Hebrew Prose* (Atlanta: Scholars Press, 1976); Gary Rendsburg, "Late Biblical Hebrew and the Date of P," *Journal of the Ancient Near Eastern Society* 12 (1980): 65–80; Ziony Zevit, "Converging Lines of Evidence Bearing on the Date of P," *Zeitschrift für die Alttestamentliche Wissenschaft* 94 (1982): 502–9; Avi Hurvitz, "The Evidence of Language in Dating the Priestly Code," *Revue Biblique* 81 (1974): 24–56; *A Linguistic Study of the Relationship Between the Priestly Source and the Book of Ezekiel,* Cahiers de la Revue Biblique (Paris: Gabalda, 1982); בין לשון ללשון (Jerusalem: Bialik Institute, 1972); "Continuity and Innovation in Biblical Hebrew—The Case of "Semantic Change" in Post-Exilic Writings," *Abr-Naharaim* Supp. 4 (1995): 1–10; "The Usage of שש and בוץ in the Bible and Its Implication for the Date of P," *Harvard Theological Review* 60 (1967): 117–21.

25. For the earlier phase, for instance Hebrew epigraphy. For the later phase, for instance Qumran Hebrew.

26. E. Blum, *Die Komposition der Vätergeschichte,* WMANT 57 (Neukirchen-Vluynx: Neukirchener Verlag, 1984); *Studien zur Komposition des Pentateuch,* BZAW 189 (Berlin: W. de Gruyter, 1990); J. Van Seters, *Abraham in History and Tradition* (New Haven, CT: Yale University Press, 1975); *In Search of History* (New Haven, CT: Yale University Press, 1983); *Prologue to History: The Yahwist as Historian in Genesis* (Louisville, KY: Westminster, John Knox Press, 1992); *The Life of Moses: The Yahwist as Historian in Exodus-Numbers* (Louisville, KY: Westminster, John Knox Press, 1994). Other works relating to the late dating of J include H. H. Schmid, *Der sogenannte Jahwist* (Zurich: Theologischer

Verlag, 1976); Rolf Rendtorff, *The Problem of the Process of Transmission in the Pentateuch,* trans. J. Scullion (Sheffield: JSOT, 1990; original German edition, *Das überlieferungsgeschichtliche Problem des Pentateuch,* BZAW 147 [Berlin: W. de Gruyter, 1977]); M. Rose, *Deuteronomist und Yahwist, Berührungspunkte beider Literaturwerke,* ATANT 67 (Zurich: 1981). For bibliographies and analyses, see David M. Carr, "Controversy and Convergence in Recent Studies of the Formation of the Pentateuch," *Religious Studies Review* 23 (1997): 22–31; Albert de Pury, "Yahwist ('J') Source," *Anchor Bible Dictionary* 6 (1992): 1016–20; E. W. Nicholson, "The Pentateuch in Recent Research: A Time for Caution," in J. A. Emerton, ed., *Congress Volume: Leuven, 1989,* Supplements to *Vetus Testamentum* 43 (1991): 10–21; Thomas B. Dozeman, "The Institutional Setting of the Late Formation of the Pentateuch in the Work of John Van Seters," *Society of Biblical Literature: 1991 Seminar Papers,* pp. 253–64; and the group of discussions in *JSOT* 3 (1977).

27. Werner H. Schmidt notes this matter of convergence, emphasizing that the doublets and the changes of divine names coincide. See his critique of the late dating of J in "A Theologian of the Solomonic Era? A Plea for the Yahwist," in *Studies in the Period of Solomon and David* (Winona Lake, IN: Eisenbrauns, 1982), pp. 55–73.
28. Baruch Halpern, *The First Historians* (San Francisco: Harper & Row, 1988).
29. Van Seters, *Prologue to History,* p. 161.
30. Van Seters, *Prologue to History,* p. 293. I commented on Van Seters's mistaken claims concerning whether the E source uses "Elohim alone" in a review article. He responded by referring to the well-known occurrences of the combined "YHWH Elohim" in Genesis 2, which is a separate issue. As is well known to anyone who has worked on this problem, this phrase combining the two terms for the deity occurs in concentration only here in Genesis 2–3 out of the entire Torah. Such a concentration, occurring precisely at the juncture where two sources overlap for the first time in the Torah, is presumably the work of the Redactor, thus softening the effect of the change of divine name at that juncture. (This is pointed out here in "About the Translation," p. 63.) It sheds no light at all on the question of J's nonuse of the word *'ĕlōhîm* in narration.

    My review article appeared in *Bible Review* 9 (1993): 12–16. A subsequent exchange between Van Seters and me appeared in *Bible Review* 10 (1994): 40–44.
31. Friedman, *The Exile and Biblical Narrative,* pp. 119–32; "Sacred History and Theology: The Redaction of Torah," in Richard Elliott Friedman, ed., *The Creation of Sacred Literature* (Berkeley, CA: University of California Press, 1980), pp. 1–2, 25–28; *Who Wrote the Bible?* (New York: Simon & Schuster, Summit, 1987), pp. 234–41.

32. J = Gen 28:11a, 13–16, 19. E = 28:11b–12, 17–18, 20–22.
33. J = Gen 28:19. E = 28:17–18, 20–21; cf. 31:13; 35:1, 6–7.
34. Sean McEvenue, "The Speaker(s) in Ex 1–15," in *Biblische Theologie und gesellschaftlicher Wandel: fur Norbert Lohfink,* eds. G. Braulik, W. Gross, and S. McEvenue (Freiburg im Breisgau: Herder, 1993).
35. Rolf Rendtorff, *The Problem of the Process of Transmission in the Pentateuch,* p. 41.
36. R. North, "Can Geography Save J from Rendtorff?" *Biblica* 63 (1982): 47–55.
37. Van Seters, *Bible Review* 10 (1994): 43.
38. Alternatively, one might also imagine the J author saying *"the* ark" even if the ark has not been mentioned before. The definite article might have seemed perfectly appropriate to an audience to whom the story of the ark and flood was already famous. I used "the ark" in my translation in *Who Wrote the Bible?* and "an ark" in the translation that appears in this book. Either is suitable, and neither is affected by Van Seters's objection.
39. Van Seters, *Prologue to History,* p. 155f.
40. See also Nicholson's criticism of Van Seters's treatment of the *Catalogue of Women* in "The Pentateuch in Recent Research," pp. 16f.
41. Van Seters, *Prologue to History,* p. 182.
42. Van Seters, *Prologue to History,* p. 270.
43. Equally perplexing is Van Seters's reading of a verse from Leviticus that plays an important part in an argument he makes about the biblical promises to the patriarchs. Lev 26:42 says:

> "then I will remember my covenant with Jacob, and also my covenant with Isaac, and also my covenant with Abraham I will remember, and I will remember the land." (Van Seters's translation)

Van Seters says:

> The first peculiarity about this text is the remarkable reverse order that is not easily explained as original. The second problem is the mention of three covenants when Genesis speaks only of one, that with Abraham.

In the present state of literary studies of the Bible it is hard to believe that Van Seters cannot imagine a writer referring to Jacob and then Isaac and then Abraham instead of in chronological order. And it is hard to believe that he can only imagine that this passage, by repeating the word "covenant," must be talking about three separate covenants. When Martin Luther King, Jr. kept repeating the phrase "I have a dream," did Van Seters think that he was referring to many different dreams? Yet from this reasoning Van Seters constructs a complex history of how this passage was written and then rewritten.

44. David Noel Freedman, "Pentateuch," *Interpreter's Dictionary of the Bible* (1962) 3: 711–27; also published in Freedman, *Divine Commitment and Human Obligation* (Grand Rapids, MI: Eerdmans, 1997), pp. 99–132.

45. Menahem Haran, *Temples and Temple Service in Ancient Israel* (New York: Oxford University Press, 1978); Yehezkel Kaufmann, *The Religion of Israel,* trans. and ed. Moshe Greenberg (Chicago: University of Chicago Press, 1960), abridged from the Hebrew edition, 1937.

46. Van Seters, *Prologue to History,* p. 281.

47. Van Seters, *Prologue to History,* p. 205.

48. Van Seters too readily dismisses other scholars' arguments with remarks such as "hardly convincing," "spurious," "rather strained," "confused," "flawed from the start," "argument becomes quite forced," "confuses the issue badly," "the whole proposal becomes so implausible and complex," "hopelessly complicated and scarcely convincing," "has greatly muddled the issue," "[his] attempt . . . seems arbitrary," "This is inconvenient for his view, so he eliminates all these references . . . ," "a little desperate." Van Seters does his work a disservice with this kind of dismissive pronouncement in the place of direct response.

49. Van Seters, *Prologue to History,* p. 44.

50. Van Seters, *Prologue to History,* p. 7.

51. Richard Elliott Friedman, ed. *The Poet and the Historian,* Harvard Semitic Studies 26 (Atlanta: Scholars Press, 1983).

52. "The Prophet and the Historian: The Acquisition of Historical Information from Literary Sources," in *The Poet and the Historian,* p. 1. Ernest Nicholson, as well, has made this point and has noted other basic distinctions between biblical accounts and Greek history writing. Nicholson, "Story and History in the Old Testament," in *Language, Theology, and the Bible: Essays in Honor of James Barr* (Oxford: 1994), pp. 135–50.

Also by Richard Elliott Friedman

Who Wrote the Bible?

This highly acclaimed bestseller brings the words of the Bible and the people who wrote it vividly to life in a fascinating work by one of our foremost Bible scholars. In an analysis that reads as compellingly as a detective story, Richard Elliott Friedman focuses on the central books of the Old Testament—Genesis, Exodus, Leviticus, Numbers, and Deuteronomy—and makes a persuasive argument for the identities of their authors. *Who Wrote the Bible?* enriches our understanding of the Bible as literature, as history, and as sacred text, and is indispensable for anyone who loves and reads the Good Book.

"Provocative . . . Friedman has gone much further than other scholars in analyzing the identity of the biblical authors." —*U.S. News and World Report*

"Thought-provoking. . . . Of interest to anyone who, aware of the unevenness and problems in the biblical text, seeks a sympathetic and perceptive guide."
—*New York Times Book Review*

"*Who Wrote the Bible?* is a fascinating and brilliant book. . . . It is full of new insights and fresh discoveries. I have spent much of my lifetime reading books about the Bible and must confess that I do not remember another that I could not lay aside unfinished."
—Frank Moore Cross, Hancock Professor of Hebrew and Other Oriental Languages, Harvard University

The Hidden Face of God

Named by *Publishers Weekly* as a "Best Book of the Year," this important work (originally published as *The Disappearance of God*) asks the question: Why does the God who is known through miracles and direct interaction at the beginning of the Bible gradually become hidden, leaving humans on their own by the Bible's end? Richard Elliott Friedman explores this and other mysteries in a bold work that encompasses religion, science, history, cosmology, and mysticism. *The Hidden Face of God* is wonderfully readable and illuminating, with a conclusion that offers real hope in a time of spiritual longing.

"Nothing less than the masterpiece for this year and years to come."
—David Noel Freedman, general editor of *The Anchor Bible*

"Lively and engaging . . . grounded in solid scholarship and passionately argued."
—Harold S. Kushner

"Friedman's biblical analysis is brilliant. An elegant and learned reflection on one of the central mysteries of the Bible and of modern life."—*Bible Review*